Quantum Physics:

The Nodal Theory

by

Hector C. Parr

Copyright © Hector C. Parr 2006

The rights of Hector C. Parr to be identified as the author of this work have been asserted by him.

First published as pages on the World Wide Web, 2002.

Things on a very small scale behave like nothing that you have any direct experience about. They do not behave like waves, they do not behave like particles, they do not behave like clouds, or billiard balls, or weights on springs, or like anything that you have ever seen.

> Richard Feynman,
> *Lectures on Physics, Book III*

In search of an explanation for the observed temporal asymmetries -- for the observed difference between the past and the future, in effect -- people unwittingly apply different standards with respect to the two temporal directions. The result is that the asymmetry they get out is just the asymmetry they put in. .... It is simply that temporal asymmetry is so deeply ingrained in our ways of thinking about the world that it is very difficult indeed to spot these asymmetric presuppositions.

> Huw Price,
> *Time's Arrow & Archimedes' Point*

Preface

The behaviour of atomic and sub-atomic particles is well described by Quantum Theory, and scientists engaged in research in this field can use the theory to make predictions of quite remarkable accuracy. But any attempt to understand the philosophy underlying the theory leads to puzzling contradictions.

During the twentieth century, several physicists with a thorough understanding of the theory have tried to resolve these conceptual difficulties, but no proposed solution seems wholly satisfactory, and each contains elements which are barely credible.

This book examines the writings of some of these scientist-philosophers, and makes one more attempt to get to the root of the problem. The author believes that a failure to understand the true nature of *Time* lies at the heart of the difficulties, and he makes one new suggestion, which he calls the Nodal Hypothesis, in an attempt to resolve them. The book shows that several deductions from this hypothesis agree well with observation, and it outlines the direction in which future research may be able to confirm that it is indeed a step in the right direction.

These pages should prove of value to everyone who is fascinated by these matters, and particularly to students who are tackling the formal discipline of Quantum Mechanics, but find little time to consider the philosophical implications. The writer is well aware that some of his arguments are not pursued to their conclusions, and suggests several areas in which younger minds should be able to advance them further.

Contents

		Page
Preface		5
Chapter 1	Preview	9
Chapter 2	Time	23
Chapter 3	Probability and Quantum Amplitudes	43
Chapter 4	The Problems	59
Chapter 5	The Nodal Viewpoint	73
Chapter 6	Interference	87
Chapter 7	Momentum	97
Chapter 8	Advanced and Retarded Radiation	105
Chapter 9	The EPR Problem	115
Chapter 10	The Special Theory of Relativity	125
Chapter 11	The Nodal Wave Function	135
Chapter 12	Conclusions	149
Appendix 1	Nodal Wave Function of a Free Particle	162
Appendix 2	The Doppler Wavelength Shift	164
Appendix 3	Quantum Computing	167
Bibliography		179
Index		181

Chapter 1: Preview

1.01 One of the great success stories of the twentieth century is to be found in the realm of sub-atomic physics. The emerging techniques of quantum mechanics enabled physicists to probe ever more deeply into the secret world of the fundamental particles which make up all the material and radiation in the universe, and these techniques have given us control over some of nature's most obscure processes. Among the inventions which quantum theory has made possible are the transistor and the laser, on which today's electronic revolution is founded, and these in turn have led to the development of that vast range of products for the consumer market with which we are all familiar.

1.02 But while we enter the twenty-first century with unprecedented skills in manipulating the natural world, we have no generally agreed understanding of why our methods work. Most scientists engaged in research or development in this field will seldom need to think about the philosophical basis for the techniques they use, but those who do are still confronted with problems and paradoxes of a type with which the human mind has never before had to grapple. Several imaginative solutions to these problems have been suggested, but each involves ideas which are difficult to accept, and not one of the proposed interpretations seems wholly satisfactory.

1.03 The greatest thinkers of the twentieth century, among whom we include Albert Einstein, Werner Heisenberg and Richard Feynman, always wrote and spoke with great clarity. They were particular over their choice of words, and careful not to let their fascination with the new phenomena lead them to exaggerate the strangeness of the world they were exploring. But unfortunately some of their colleagues were not blessed with the same clear vision or lucidity of expression, and their writing sometimes lacks the precision which is essential if they are to present the mysteries of the quantum world without leaving conceptual gaps into which others can stumble. The problems are so baffling that it would not be fair to blame our failure to resolve them on any individuals, but it does seem likely that the rather slack writing of some commentators, and their tendency to sensationalise the unexpected nature of the effects they describe, have retarded the search for an acceptable philosophy.

1.04 Soon after the start of the twentieth century it was realised that, in the atomic domain, things which had been assumed to be *particles*, such as electrons and atoms, often manifested themselves as *waves*, and conversely things like *light*, which we had taken to be electromagnetic waves, sometimes behaved as if they consisted of particles, which were given the name "photons". This strange duality, and the real nature of these waves and particles, presented a problem which seemed to become more rather than less obscure as our knowledge increased, and now a century later, is far from being fully understood. Much of the discussion and disagreement has centred on the nature and effects of *measurement* in the quantum world, and the so-called "collapse of the wave function" which appears to occur whenever a measurement or an observation is made of a quantum system. It is here in particular that careless choice of words can exacerbate a problem rather than illuminate it. A particle cannot be observed or measured without disturbing it to some degree. The same principle applies when making macroscopic measurements, for instance in electrical circuits where any measurement of the current flowing through a circuit element necessarily affects the voltage across it, and *vice versa*, but whereas these effects can be minimised to any extent by careful design of experiments, in the quantum world there is a natural lower limit to the level of disturbance that must be tolerated because of the *quantised* nature of energy. The apparatus we set up to measure or observe something in the atomic world necessarily affects the quantity we are measuring to a degree we cannot predict or reduce.

1.05 The Institute of Physics which Niels Bohr established at Copenhagen in the early 1920's became the chief centre at which these matters were discussed, and many of the greatest physicists of the day visited Bohr there and contributed to the deliberations. The "Copenhagen Interpretation" which Bohr and his friends thrashed out in these early days is still regarded by many observers as the authoritative explanation of the quantum paradoxes, but this interpretation leaves many questions unanswered, and indeed seems to mean something different to almost every writer who describes it. Bohr appeared to insist that we should talk about the behaviour of waves or particles only in relation to particular types of observation or experiment, and that it was meaningless to discuss their "reality" in general terms. But on this matter Bohr's view was strongly opposed by Einstein. Throughout his life Einstein maintained a belief in the "real" existence of particles in all circumstances, and the doubts which quantum theory had to throw on this reality, he maintained, indicated that the theory was wrong, or as Einstein expressed it, charitably, was "incomplete".

1.06 Niels Bohr in his writings made it clear that in talking about an "observer" causing the collapse of a wave function he was referring merely to the *apparatus* involved, and attached no importance to whether a human was actually aware that an observation had been made. The collapse was brought about by "irreversible amplification effects" such as the production of marks on a photographic plate, or the "building of a water drop around an ion in a cloud-chamber". And Werner Heisenberg, writing in 1958, explains that, in its later form, the Copenhagen Interpretation envisaged this wave function to be partly subjective, containing information about *our knowledge* of a system as well as information about the system itself.

> This probability function represents a mixture of two things, partly a fact and partly our knowledge of a fact. ...It should be emphasised, however, that the probability function does not in itself represent a course of events in the course of time. It represents a tendency for events and our knowledge of events. (*The Copenhagen Interpretation of Quantum Theory*, ch.3)

1.07 But this uncontentious portrayal of the wave function (Heisenberg's "probability function"), as it was passed from one commentator to another, became more and more sensational and outlandish. Paul Davies is one of the more rational writers on scientific topics, and yet he could write in 1987,

> So long as a quantum system is not observed, its wave function evolves deterministically. In fact, it obeys a differential equation known as the Schrodinger equation (or a generalisation thereof). On the other hand, when the system is inspected by an external observer, the wave function suddenly jumps, in flagrant violation of Schrodinger's equation. The

system is therefore capable of changing with time in two completely different ways:one when nobody is looking and one when it is being observed. (*The Cosmic Blueprint*, p.168).

John Wheeler goes further, and appears to attribute supernatural powers to the human observer:

> The quantum principle shows that there is a sense in which what the observer will do in the future defines what happens in the past -- even in a past so remote that life did not then exist, and shows even more, that 'observership' is a prerequisite for any useful version of 'reality' (Quoted by Paul Davies in *Other Worlds*, p.126).

And according to Davies, Eugene Wigner takes us yet further along this road;

> It is not enough to equip the laboratory with complicated automatic recording devices, video cameras and the like. Unless somebody actually looks to see where the pointer is on the meter (or actually watches the video record), the quantum state will remain in limbo. (*The Ghost in the Atom*, p.31).

Bohr would certainly not have approved of some ideas which have been attributed to him by more recent popularisers of quantum physics. We read,

> The strangest thing about the standard Copenhagen interpretation of the quantum world is that it is the act of observing a system that forces it to select one of its options, which then becomes real. ... What's worse, as soon as we stop looking at the electron, or whatever we are looking at, it immediately splits up into a new array of ghost particles, each pursuing their own path of probabilities through the quantum world. Nothing is real unless we look at it, and it ceases to be real as soon as we stop looking. (*In Search of Schrodinger's Cat*, John Gribbin, 1984).

1.08 Such provocative writing may help to sell popular science books, but has done nothing to advance the understanding of quantum physics. Contrast it with Bohr's cautious approach to the problem of observing quantum phenomena;

> ... I warned especially against phrases, often found in the physical

literature, such as 'disturbing of phenomena by observation' or 'creating physical attributes to atomic objects by measurements'. Such phrases, which may serve to remind of the apparent paradoxes in quantum theory, are at the same time apt to cause confusion, since words like 'phenomena' and 'observations', just as 'attributes' and 'measurements', are used in a way hardly compatible with common language and practical definition. (*Discussions with Einstein*, 1949)

1.09 But the ultimate accolade for eccentricity must go to Hugh Everett. Here is Davies' explanation of Everett's "many worlds" theory:

> ... if a quantum system is in a superposition of, say, *n* states, then, on measurement, the universe will split into *n* copies. In most cases *n* is infinite. Hence we must accept that there are actually an infinity of "parallel worlds" co-existing alongside the one we see at any instant. Moreover, there are an infinity of individuals, more or less identical with each of us, inhabiting these worlds. (*The Ghost in the Atom*, p.35)

Paul Davies himself, however, can bring us down to earth again, for just a few pages beyond the passage quoted above, in which he attributes different behaviour to a system according to whether or not someone "is looking", he gives us this much more reasonable assessment:

> The wave function represents not how the system is, but what we know about the system. Once this fact is appreciated, the collapse of the wave function is no longer so mysterious, because when we make a measurement of a quantum system our knowledge of the system changes. (*The Cosmic Blueprint*, p.172).

1.10 The belief held by some physicists that their own conscious awareness of a measurement somehow affects the particles in the physical world which have been observed, appears to have arisen only because of careless (or sensational) statements about the *measuring apparatus* having such an effect on those particles. It is then just a short step to believing that the wave function which collapses when it knows a measurement has been made, is itself a part of the system being observed, rather than a description of the physicists' own thought processes, and that its collapse is a physical change in the system brought about by the observer's changed perception of it.

1.11 Further examples of bad reporting can be found in the many descriptions of the "Heisenberg Uncertainty Principle". Heisenberg showed that, if we try to measure simultaneously the position and the velocity of a particle, then there is an essential uncertainty in one or both of our measurements. It is more convenient to express this fact not in terms of the particle's *velocity*, but of its *momentum*, which is found by multiplying the velocity by the particle's *mass*. Then Heisenberg tells us that the uncertainty in the position multiplied by the uncertainty in momentum must always exceed h, Planck's constant. (This small number plays an important part in Quantum Mechanics, and its value is known accurately.) Other writers tell us less formally that "it is impossible to know both the position and the momentum of a particle". But such a statement is not nearly good enough; under what circumstances does it apply? Does it refer to the position and momentum at some time in the *past*, in the *present*, or the *future*? Does it concern the values of these variables at the moment a particle is *emitted*, at the moment it is *detected*, or at some time *in between*? The writers do all agree that there is no fundamental restriction on the accuracy with which we can measure the time and position of the *emission* of a particle, or of its *detection* by a measuring device, provided we do not at the same time try to find out anything about its *momentum* or its *energy*. But for a particle in free space, where we can assume it moves in a straight line at constant speed, if we know the place and time at both ends of its track, we can calculate the particle's velocity (to whatever degree of accuracy our methods allow) by simply dividing the distance travelled by the time taken. This gives us the momentum, again with a degree of accuracy limited only by our experimental techniques, and not by fundamental restrictions of the uncertainty principle. As usual, Richard Feynman can be trusted to describe the situation clearly. He writes,

> It is quite true that we can receive a particle, and on reception determine what its position is and what its momentum would have had to have been to have gotten there. That is true, but that is not what the uncertainty relation refers to. This equation refers to the predictability of a situation, not remarks about the past. (*Lectures on Physics III*, 1965, Addison-Wesley).

We see that Heisenberg's rule may not apply to *past* trajectories of a particle. But careful examination of situations in which the position-momentum uncertainty relationship *does* apply is revealing. The simplest case occurs when a collimated beam of particles is projected onto an opaque screen with a small hole in it, and they appear to be deflected from a straight course by diffraction. Those particles which pass through the hole have a well-defined position here, and this entails a random change of momentum in accordance with the Heisenberg rule, so that we cannot predict the subsequent direction of a particle's trajectory. Notice that in this case we know the particle's position not because we have observed it there, but by "default" or "selection". Most of the incident particles do not pass through the hole, for they collide with the screen, and our experiment does not reveal the positions of these collisions. But the small selection of particles which do go through the hole have a known position simply because they have *not* collided with the screen. It appears, therefore, that uncertainty arises only when we are interested in the position and the momentum of a particle *between* two collisions, or at one collision without using information we could gain from its previous one. If we know the time and position of a consecutive pair of *past* collisions we may well be able to determine the momentum between the collisions without the restrictions of the uncertainty relation. This is not made clear in many of the written accounts of the principle.

1.12 A crucial fact that is often forgotten is that a particle can have no direct effect on another particle unless it *collides* with it, or unless a third particle such as a photon travels from one to the other. (Notice that, when two particles are described classically as attracting or repelling one another, a quantum description can always be given in which the influence is transmitted via third particles, such as photons or gravitons.) It follows that a particle can have *no influence whatsoever on anything else* between one collision (or emission) and the next one. Any apparatus we set up to observe or to measure a particle must necessarily involve a new collision with that particle. The only information we can gain in the absence of such a collision must be *by default*. This simple truth is the basis for the new attempt which the writer makes on these pages to resolve the difficulties surrounding quantum mechanics without undue threat to credibility; he calls this approach the Nodal Interpretation, and will present its main features in the following chapters.

1.13 Let us take as a starting point the strange fact that, because you can know nothing at all about a particle between one collision and the next, you can assert anything you like about its trajectory and no-one can prove you wrong. What the Nodal theory asserts is that a particle *does not exist* between collisions. The universe consists not of particles, but just of those events which we have come to regard as collisions between particles, together with the wave functions which connect these collisions, and pass information between them. We call these events *nodes*, and we shall examine in later chapters what properties they must possess if we are to explain all quantum phenomena in terms of nodes alone. The particle does not travel from node to node, but there is a sense in which *information* connects the nodes to each other, and some sort of *wave function* carries this information. In chapter 11 we examine in detail the nature of this wave function, and discover in what respects it resembles the wave function of traditional quantum theory.

1.14 Just as a foretaste, consider how this theory deals with the well-known two-slit experiment, in which a beam of monochromatic light passes through two slits in an opaque screen, and is seen to produce an *interference* pattern, consisting of alternate dark and light bands, when the divided beam falls onto a photographic plate. This was easily explained by nineteenth century physics, for light was then believed to consist of electromagnetic waves, and because the two routes to any particular point on the plate can differ slightly in length, the two waves may be in phase or out of phase; they are said to "interfere". Where they are in phase they reinforce each other, and a bright band results, but where they are out of phase they can neutralise each other and give a dark band. What is far more difficult to explain is that the experiment can be repeated with a beam of *particles* such as electrons instead of light rays, and is found to give a similar pattern of light and dark bands. How can this arise, when interference is essentially a property of waves, and not of *particles*? Feynman described the phenomenon as being at "... the heart of quantum mechanics. In reality, it contains the only mystery" (*The Feynman Lectures III*, 1965). When a single particle traverses the apparatus it must "know" it has to avoid certain areas of the screen, namely those in the dark fringes, but only if *both* slits are open. If one slit is closed no interference pattern is produced; so closing one slit can *increase* the number of particles arriving at points within the dark areas. We ask how the particle can know whether the slit through which it has *not* passed is open, or whether, perhaps, the particle splits in half and passes through *both* slits. But a simple consideration of the masses of the two parts, and their corresponding wavelengths, shows that this last suggestion cannot be the true explanation, and it has become fashionable for popular writers on quantum theory to talk about particles being "in two or more places at the same time". This unsatisfactory description was put on a more formal footing in the early 1930's when Paul Dirac wrote about the "Principle of Superposition of States", and formulated an ingenious mathematical treatment of the idea. He describes the situation in which a particle apparently passes through two slits at the same time as a "superposition" of the two separate states in which it passes through one or other of the slits. Dirac writes:

> The nature of the relationships which the superposition principle requires to exist between the states of any system is of a kind that cannot be explained in terms of familiar physical concepts. .... The intermediate character of the state formed by superposition thus expresses itself through the *probability* of a particular result for an observation being intermediate between the corresponding probabilities for the original states, not through the *result itself* being intermediate between the corresponding results for the original states. (*The Principles of Quantum Mechanics,* Pp.12-13).

1.15 Dirac's thorough treatment, and the special notation he developed to represent quantum states and their superpositions, have formed the basis for the mathematics which practical workers in the field of particle physics have used with notable success throughout most of the twentieth century. However, they provide little help for those wanting a clear picture of the natural processes lying behind the mathematics, and it is hoped the Nodal theory might help to fill this gap. Briefly, the new theory denies that a particle exists at all while traversing the "two-slit" apparatus, so it passes through *neither* slit. We do not need to believe it can be "in two places at once". A full description of our analysis of this experiment will be found in Chapter 6.

1.16 There is one other principle to which some writers on quantum mechanics pay lip-service, but fail to pursue to its conclusions, and which we must develop to understand fully the Nodal theory. It is known that all quantum processes, when considered as the interactions of *particles* (with one trivial exception) are *time symmetric*. If time were considered to "flow" in the opposite direction, all fundamental processes would still be in accordance with the rules of quantum mechanics; we would not be able to detect, by watching the behaviour of sub-atomic systems, which way time was flowing. (This would not be true, of course, if we were observing *macroscopic* processes, such as bombs exploding, water flowing, or cars colliding, for almost all such processes are dominated by the asymmetrical second law of thermodynamics.) It follows that any *description* of interactions on the microscopic scale should also be time symmetric, and all the popular quantum philosophies *fail* this test. The collapse of the wave function, which most of these philosophies involve, is clearly an irreversible process, and so, *a fortiori*, would be the continual splitting of the universe into more and more copies of itself, which the Everett interpretation requires. This is a simple test which we can apply to any description of quantum phenomena, and which eventually we shall apply to the Nodal theory. If our picture becomes less plausible, or if it changes in any fundamental way, when time is considered to "flow" in the opposite direction, then our description is faulty.

1.17 Again, we shall quote just one simple example of this principle of reversibility, and leave fuller discussion for later chapters. When a randomly generated stream of photons strikes a *polarising filter*, such as a sheet of Polaroid, it is found that one half of them pass through it, and one half are stopped. But if a second filter is placed behind the first, and if the optical axes of the two filters are parallel, every one of the photons which passes through the first polariser passes also through the second. If the axes of the two polarisers are not parallel, a simple formula tells us what proportion of the photons passing through the first filter pass also through the second. [If the angle between the optical axes of the two filters is q, then this proportion is known to be $\cos^2 q$.] These facts are explained, in traditional quantum theory, by assuming that each photon has its own axis of polarisation, and that when a photon strikes a polariser, this same formula tells us the *probability* that it is transmitted. Those that are transmitted are supposed to have their own axes shifted so that they then become parallel to the optical axis of the polariser through which they have passed. The very act of passing a photon through a polariser, we are told, constitutes a *measurement* of the photon's axis, so we should not be surprised that it has an instantaneous effect on that photon.

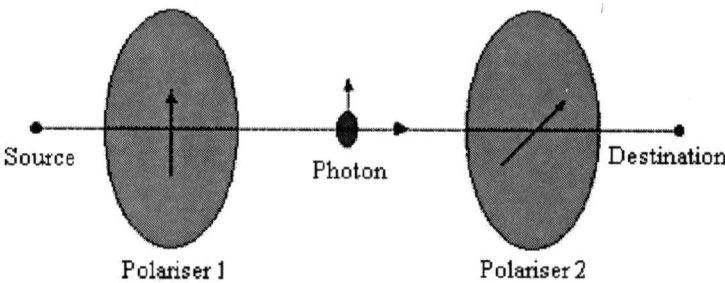

Fig. 1-1

1.18   Let us consider the possible outcomes when we try to pass a stream of photons through two such filters with their axes at some definite angle, say 60°, as illustrated here. One half of them are stopped by the first filter, and three quarters of the remainder are stopped by the second. If we concentrate on a single photon which passes through both, let us imagine what it looks like when *between* the two filters. Since it has just passed through the filter on the left, its axis must be vertical, as shown. We might prepare a little "movie" of the process, to show our students what we think happens (Fig.1-1). But suppose we accidently show the movie *in reverse*. The photon firstly passes through the *right hand filter*, and so its axis should then be at an angle of 60° as shown in Fig.1-2, and yet we can see clearly that it is not, it is vertical.

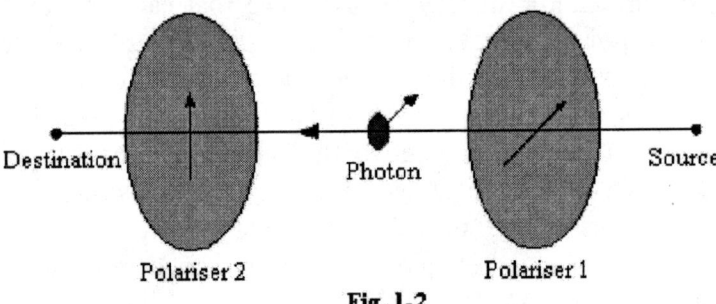

Fig. 1-2

So by looking at our movie we can tell which way time is "flowing", in breach of our principle of reversibility. In the nodal picture this problem disappears, for the photon *does not pass through the filter* at all; it exists only at its source and at its destination, on the right or left as the case may be, and not at any points in between. Note carefully that a photon which is *not* transmitted by a polariser *does* have a collision there, and so does exist as a node at that point, but we are considering only the photons which have a collision at neither polariser. Of course the presence of the polariser affects the future behaviour of the photon, and we must take it into account in calculating the relevant probabilities. But the Nodal theory claims that nothing physically significant passes through the two polarisers, and so no meaning can be assigned to the orientation of the particle's axis there, and the apparent time asymmetry is seen not to exist. It may be argued that, even from the Nodal viewpoint, the wave function must pass through the two filters. But the wave function is no more than the information which is exchanged between nodes, and it is fundamentally impossible to gain access to this information except at the nodes themselves, or by introducing extra nodes, which will change the very information we are hoping to observe. The wave function must take cognizance of any polarisers in its path, but we can observe only the nodes, and the behaviour we see does not depend upon the direction in which we imagine time to be flowing. This is the viewpoint which will be developed throughout the following chapters.

1.19 The above discussion illustrates one of the greatest difficulties in understanding quantum processes, that of untangling them from our deep-rooted intuitions concerning the flow of *time* and the true nature of *past* and *future*, intuitions which are firmly ingrained in our thinking, but in several respects are false. We devote the next chapter of the book to this question.

Chapter 2: Time

2.01 We hope to show in this chapter that the natural impressions we all possess of the true nature of Time are illusory and confusing. It is one thing to expose these false impressions by careful argument, and several books have been published in recent years suggesting that many thinkers, both physicists and philosophers, are now facing up to the problems posed by such prejudices. But it is much more difficult to purge them from our thinking, for they can creep back insidiously and upset our reasoning in subtle ways.

2.02 It seems to us that we live in a three-dimensional world of space, and that the material world which exists in this space *changes* from moment to moment as time advances. But surprisingly, all the mathematical equations and relationships which describe this world contain nothing corresponding to the idea of *change*. The basic units with which the fundamental sciences deal are not *things* in space, but *events* in space-time. The equations describing the motion of bodies, whether in everyday life, in atomic phenomena, or in behaviour on the large scale requiring the techniques of relativity theory, all involve the variable t as well as x, y and z, and these four variables occur on equal terms. They describe, in fact, a *four--dimensional* world, and one which is essentially static. The idea of change is not represented in this description by anything moving or altering, but simply by the different patterns of events which are found corresponding to different values of the t co-ordinate, just as events at different places are represented by points with different values of x, y or z.

2.03 We are unable to visualise four dimensions, but for purposes of illustration we can often dispense with one of the space dimensions, and the z co-ordinate can be omitted. We may then represent events on a perspective drawing with the x and y axes in the horizontal plane and the t axis drawn vertically upwards. The life-history of a small body or a particle of matter is represented by a *line* which we call its "world-line", showing its continued existence throughout a period of time, and its movement is represented by the *slope* of this line, the angle it makes with the t-axis.

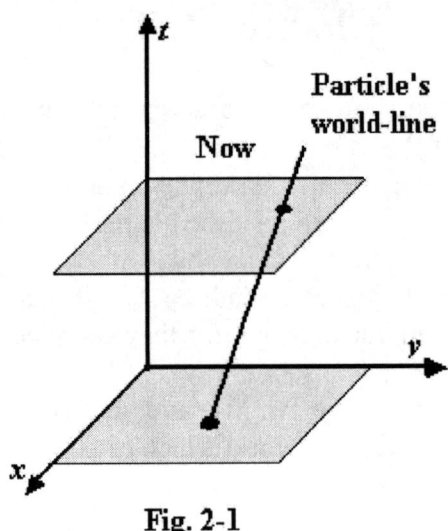

Fig. 2-1

The state of the world at the particular time we call "now" or the "present" is represented by everything on a horizontal plane drawn for a particular value of t. And we can imagine the progress of time being represented by this plane moving slowly upwards at constant speed. The strange fact that our mathematical descriptions of the world contain nothing corresponding either to the "now" plane, or to its upward movement, need not imply that our impressions are wrong; perhaps the mathematical descriptions are just incomplete.

2.04 This picture might be thought of as no more than an interesting way of representing the relationships between events, while not really indicating anything about the world itself. But this view had to be abandoned at the beginning of the twentieth century with the acceptance of Einstein's Special Relativity theory. The x, y, z, t picture was suddenly seen to present a much closer view of reality when Einstein showed that, just as the choice of directions for the x, y and z axes is arbitrary, so too is the direction of the t axis and the plane we adopt for our x-y co-ordinates. Einstein shows that both of these are dependent upon one's state of motion. If an observer A is moving relative to B, then A's x-y plane must be inclined to B's by an angle which can be calculated easily from Special Relativity theory. This shows that we must abandon the belief that the "now" plane is universal, that my "now" plane is necessarily the same as yours. Expressing this differently, Einstein has shown that the notion of "simultaneity" is meaningless. Relativity reveals many simple situations in which two events which one observer describes as simultaneous, to another observer are separated by a measurable time interval. We discuss this more fully in Chapter 10. At any particular point in his life, an observer is still free to imagine his own "now" plane if he wishes, but it is purely personal, and has no existence outside his own imagination. And if the "now" plane has no intrinsic existence in the outside world, neither can the motion of a universal "now" plane be real. The idea of a moving time is false; it exists only in the minds of individuals.

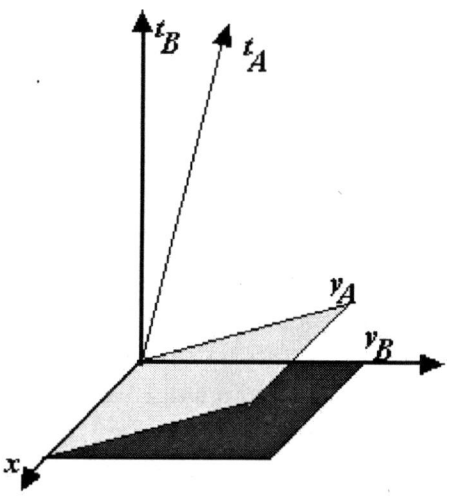

Fig. 2-2

2.05 Indeed it did not require Einstein to show that the notion of a "flowing" time is a nonsense. If the time is continually flowing or changing, as our natural intuition suggests, with respect to what does it change? One thing can change only with respect to another. Usually when we talk about *change* we mean change with respect to time itself. The rate at which a vehicle changes its position is its *speed*, measured perhaps in miles *per hour*. The rate at which an investment earns interest is measured as a percentage *per annum*. Mathematicians represent the rate of change of any quantity x by the symbol dx/dt. But time cannot be said to change with respect to *itself*; dt/dt is meaningless.

2.06   There is an even more challenging deduction from this line of thought. Not only do we think of our "now" plane as steadily advancing as we go through life, but at any moment of time it appears to divide the history of the world into two very different regions. All the events below it *have happened*. There is nothing we, or any one else, can do to alter them. But all the events above it lie in the future. Even if we can try to predict some of them, they seem to lack the degree of certainty possessed by events which lie in the past. Indeed we like to think that we ourselves can be responsible for them. We believe today's decisions and actions help to "shape the future", but it never occurs to us that we can change the past. And yet the arguments presented in this chapter show that we must now dismiss these impressions as illusory. If there is no universal "now" plane to separate the certain past from the uncertain future, there can be no intrinsic difference between events which lie in the one region from those which lie in the other. *All* events in the world's history, past present and future, must be equally certain and unalterable.

2.07   Another important element in our impressions regarding the nature of time is manifest in our intuition concerning the *irreversibility* of many processes we observe. We are amused when shown *in reverse* a moving picture of someone shuffling a new pack of cards, an egg being broken into a dish, a river flowing downhill, or an electric fire cooling down when switched off. We know that cards do not sort themselves into numerical order, broken eggs do not transform themselves into whole eggs, and electric heaters do not spontaneously heat themselves when disconnected from their power source, but we seldom consider why this should be. Some of the formal consequences of this irreversibility are seen in the science of Thermodynamics. The Second Law of Thermodynamics can be expressed in many different ways, but perhaps the simplest is to say that heat always flows from a hotter body to a cooler, and never the reverse. We know that, when we turn off a fire, the heating element cools until it has reached the same temperature as its surroundings, but it then remains constant. The laws of thermodynamics were gradually formulated during the nineteenth century as engineers strove to improve the efficiency of the newly developed steam engine. They came to realise that heat is just one form of energy, of which every body contains a certain amount, but that this energy cannot be transformed into a useful form unless temperature *differences* exist between different parts of an engine, for instance the furnace and the condenser of a steam locomotive. They devised the concept of *entropy* to describe the usefulness of the energy in a system. Thus, after an electric fire has been turned off, the total heat energy in the fire and its surroundings remains constant as the element cools, and its surroundings are warmed slightly, but the entropy gradually *increases* as the two temperatures approach each other. The Second Law of Thermodynamics tells us that the entropy of any closed system cannot decrease, and usually increases, as time advances. As the science of Thermodynamics developed, the notion of entropy helped in the description of processes involving heat transference and the working of steam engines, but it was subsequently found to be applicable to a wide range of other phenomena. Of special importance to us in our investigation of the nature of Time, as we shall see, is the application of entropy to a system of bodies moving under the influence of gravitational forces, and to the universe as a whole.

2.08 The very familiarity of processes which are essentially irreversible makes their explanation more difficult. Indeed it has only gradually come to be realised that any explanation is required, and many who have attempted to understand them have been led astray by subtly assuming that which they are attempting to prove. The difficulty of the problem was only slightly lessened when it was realised that all manifestations of the Second Law have one thing in common, namely that the later state of a system is almost always a *more probable* state than the earlier state. It is more likely that a pack of cards are in random order than that they are sorted, for there are many more random orderings than sorted ones; if the molecules of an egg lie in a basin, it is more probable that the yoke and white will be mixed together than completely separate; and if we know the average energy of all the molecules of the gas contained in a flask, it is more probable that the energies are distributed randomly than that the molecules in the left half have a greater average energy than those in the right. But in attempting to explain why the more likely situation always seems to occur at the *later* time, we are very prone to be led astray because the fact seems so obvious. And we are likely also to overlook the most paradoxical element of this question. All the fundamental laws of physics, including the laws of mechanics which describe the motions of the molecules of a gas, and which explain their behaviour not only as they move unimpeded between one collision and the next, but also as they collide with each other and with the walls of the containing vessel, these laws are all *time symmetric*. If we represent the motions of a small number of molecules as a moving picture and view this with time reversed, the molecules will continue to obey the same laws. We would be unable, by viewing such a moving picture, to determine whether it was being displayed correctly or in reversed time. The mystery which requires explanation is how the time-symmetric laws which govern the motion of the individual molecules can give rise to the time-asymmetric Second Law which describes, on a less detailed scale, these same motions.

2.09 The great nineteenth century physicist Ludwig Boltzmann (1844-1906) attempted to answer this question with his so-called H-Theorem. He knew that, when the molecules of a gas in a container were in a condition of equilibrium, corresponding to what we would today describe as maximum entropy, the statistical distribution of energies of the individual molecules took a particular form which had been derived by Clerk Maxwell (1831-1879) several years previously. Boltzmann presented a proof of the fact that any other distribution of velocities would steadily change until the equilibrium state was reached. If, for instance, the molecules in one half of a container have a higher average energy than those in the other half, this state of affairs will dissipate as time progresses and the high energy molecules collide with those of lower energy. His proof was viewed with suspicion by several physicists of the day, and a number of challenges were mounted. These all hinged upon the fact that possible states of motion could be postulated which would result in a *decrease* rather than an increase of entropy, and a movement *away* from equilibrium. Imagine, for example, a gas which was formerly in an unstable state, perhaps with a temperature gradient across it, but which has now reached equilibrium. If we suppose that each molecule has its velocity *reversed* the gas will then trace in reverse its previous history, and entropy will steadily decrease. It is irrelevant that we know of no way to bring about this reversal; the reversed motion is perfectly possible, and in fact is as likely to occur by chance as was the actual motion before reversal. Arguments of this sort show that Boltzmann's proof of the H-Theorem must contain a flaw, and as a result Boltzmann came to realise that what he had proved was not a *certainty* but rather a situation which was overwhelmingly *likely* to occur. The fact that, in this new form, the H-Theorem still did not explain how a gas evolved in a time-asymmetrical manner as a result of the time-symmetrical behaviour of its molecules, was still not fully realised by either Boltzmann or his critics, and led to almost a century of argument, much of which was of no consequence because the protagonists were themselves misled by their deep-rooted misconceptions of the irreversibility of time itself.

2.10 Only gradually was it realised that the Second Law did not result from the fundamental laws of science, for these are time-symmetric, and nor did it arise because of any statistical asymmetries. In fact it is a purely *empirical* fact, arising from the very special state of the matter in the universe at the present time. What requires explanation is not the movement of the universe towards higher entropy, towards an increasingly probable or disordered state; rather it is why entropy today is so *low*, and why the universe is at present in such an unlikely state. Almost everything of interest that happens on earth, including the sorting of cards into numerical order by humans, the manufacture of eggs by hens, and the raising of water vapour to the tops of mountain so that rivers can flow downhill, is ultimately a result of the large temperature difference which exists between the earth and the sun, a manifestation of low entropy to which we, and all living things, owe their existence. The problem thus becomes one for cosmologists rather than heat engineers, and in recent years they have begun to make some progress in solving it. They are revealing that the unsolved question is one of *gravitation* rather than of *heat*. The sun and the stars shine only because they have condensed by gravitational attraction from more tenuous and more homogeneous matter, and the question for the cosmologists is why this did not happen long ago, resulting in all the matter in the universe condensing into one enormous mass or one black hole. They are showing us that the *expansion of the universe* is an essential part of the explanation, for it clearly acts in the opposite direction to gravitational attraction. But they also tell us that the degree of homogeneity of the early universe must have been critically determined for it to have persisted for so many billions of years, and yet still not have reached equilibrium, with all its matter condensed into one mass, or at least with everything at the same temperature. If the early universe had been slightly too uniform it would not yet have condensed into galaxies and stars, while if it were insufficiently uniform the condensation would have proceeded too far, with the galaxies already collapsing into black holes. Everything depends on the nature of the big bang, which is far from being understood at the present time. For our purposes it is not necessary to pursue this further; it is sufficient if we acknowledge that the conditions soon after the big bang were such that entropy has remained far below its maximum value for some fifteen billion years, and we are witnessing the universe during an era when its gradual "running down" is producing some fascinating phenonema, including ourselves!

2.11 In the previous paragraph we talk about *movement*, things *happening*, the universe *expanding* and the stars *condensing*. Is not this all in the language of a *flowing* time, a concept which we have just described as "nonsense"? It is, indeed, but we often need to use such language because we have no other. Our language evolved long before anyone questioned whether time flowed and things happened, or before anyone thought of picturing the four dimensions of space-time. But we can often clarify our thinking by translating such language into the more accurate descriptions provided by pictures of the corresponding static structures in space-time. Let us apply this technique to a consideration of the expansion of the early universe. In our picture we must represent *time* by the distance from the centre of our diagram rather than by distance up the page, and we can then represent the whole of space at any particular time by a circle whose radius corresponds to that particular time. We thus get a very simple picture in two dimensions only, with the three dimensions of space portrayed by the one dimension of the circumference of a circle. The resulting pattern resembles the lines of latitude and longitude around the North Pole, with time represented by the meridians and the whole of space at any particular time represented by one of the circles of latitude.

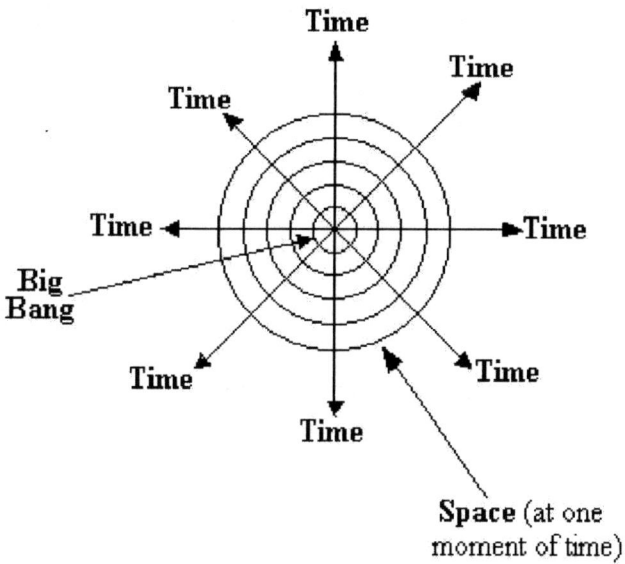

Fig. 2-3

The world lines of particles of matter and radiation then become lines radiating from the pole, always extending in a generally outward direction, but deviating slightly from the lines of longitude to show their intrinsic motion, and meeting each other where our picture needs to represent their collisions. The homogeneity of the early universe, which was so finely tuned to give us the sort of world we now find, is represented on our picture by "boundary conditions" which these world lines must satisfy close to our "North Pole". A familiar and useful analogy is provided by the electric field around a small charged body; the distribution of charge on its surface provides the boundary conditions which determine the electric field close to the surface, and this in turn determines uniquely the direction and strength of field throughout the whole space influenced by the charged body. In the same way, the whole future of the universe is constrained by the boundary conditions at the big bang. That future is not *uniquely* determined because of various types of uncertainty which we shall discuss later, but the general macroscopic picture is dominated by the very low entropy represented by the homogeneity of the world-line distribution near the pole, and the Second Law is our description of the increasing entropy at greater distances (i.e. later times), and particularly by the clumping together of world-lines under the influence of gravity.

2.12 We must examine some further consequences of the Second Law. We have seen that the law does not apply to micro-systems consisting perhaps of a small number of molecules, or the collisions of a small number of atomic or nuclear particles, for reactions of this type are always time-symmetric; if they had a degree of asymmetry in the very early universe they have long ago worked this out of their system. The Second Law does apply to macro-systems such as packs of cards, eggs, and flasks full of gas, where we are not interested in the behaviour of individual molecules, but only in larger scale characteristics such as temperature or pressure. It was pointed out that the Second Law can be pictured informally as a progression to states of increasing *probability*, i.e. towards macro-states which would arise from increasing numbers of possible micro-states. Another useful picture is presented when we realise that processes which proceed to higher values of entropy always contain some element of *dissipation*, or *dispersion*. A new pack of cards is arranged in perfect order, and as shuffling progresses this order is gradually dispersed through the pack; at first there will still be groups of cards which are in numerical order, but these groups become smaller, and more numerous, until they are completely dissipated throughout the pack. When a hot body cools to the temperature of its surroundings, its heat energy is gradually dissipated throughout a greater volume of matter. And almost all the processes we see around us are dissipative in some way. Those that seem to go against this rule, such as heat engines, refrigerators, or the evolution of life itself, will all be found on examination to involve a greater degree of dissipation into their environment, a greater overall increase in entropy, than the decrease which they themselves represent. Entropy always increases in any self-contained system. The answer to the question of why some parts of the universe are still far from equilibrium is that the organisation they display is left over, as it were, from the highly ordered state of the early universe. They are still linked to the boundary conditions existing just after the big bang.

2.13 An important example of the irreversible dissipation of such structures is shown whenever a system generates a *record* of itself, whereby a picture or copy is made of some of its characteristics, for elements of the structure of this system are duplicated in the record. This is dissipation, albeit in a minimal form; whereas there was formerly just one instance of these elements there are now two or more. Regarding the process of record-making in this way, it is seen how the Second Law provides a simple explanation of the fact that records can exist only of *past* events. Consider, for example, the making of a number of copies of a document in a photo-copier. The original and all the copies contain no more information than the original itself, but this information is distributed more widely after the copies have been made, which is thus an irreversible process.

2.14 A most significant example of record formation exists in the human brain, the laying down of *memories*. As we shall see in later chapters, the existence of memories of past events, and the impossibility of having memories of the *future*, plays a large part in the muddle which has characterised attempts throughout the twentieth century to understand quantum behaviour. So long as we believe that the asymmetry of record formation and memory formation is just obvious, and requires no explanation, just so long will the muddle remain unresolved.

2.15 Thus we see that our common perception of the nature of Time carries with it *four* features which are false, and which are prone to mislead us in our thinking about the real world:

(i) We believe that the particular moment of time which we call "now" (and which we will denote by t_0) has an objective significance independent of our own minds, and that the whole universe shares this same value of t_0.

(ii) The value of t_0 seems constantly to *increase* at a constant rate (whatever that may mean).

(iii) The time t_0 appears to divide the history of the universe into two parts, events occurring at times t for which $t < t_0$ appearing to have a degree of certainty not enjoyed by those for which $t > t_0$.

(iv) We accept the Second Law of Thermodynamics believing it to require no justification. In other words it seems natural that entropy should increase, and that systems become *more* disordered as the value of t increases, and never the reverse.

2.16 In the remainder of this chapter we shall try to determine why we have these false impressions, and what we can do to prevent them from leading us astray in our attempts to understand the fundamental workings of the universe, and in particular the interpretation of quantum phenomena.

2.17 Our belief in an objective and universal "now" is not difficult to understand. Every event has a position in space and a position in time. And every thought we have is an event; it is situated in our brain, and it occurs at some particular point in time. At the same moment that we experience a thought there are usually other events going on around us, and it is natural to suppose that more remote events of which we are not aware, even events at the other end of the universe, will be occurring at this same time. This present moment has a special significance for us. It is the time when things happen to us, and the moment at which we are acting; indeed it seems to us to be the last possible instant at which we can affect events lying in that moment's future. So it is natural for us to suppose that this particular t value is special in its own right, the time at which everyone else is being affected or is making decisions, the moment which has been reached by the whole universe.

2.18 More difficult to explain is the overriding impression we have that this "now" is constantly advancing, that time constantly changes or flows. This sensation is certainly related to our possession of *memory*; if we had no memory, not even of the events which immediately preceded the present moment, we could have no feeling of the progression of time. It is the memories we have of our past life that make plausible our belief that we have lived that life up to the present moment, re-inforced by all the external records we have of that past life, and we hope and believe we shall continue living it in the future, collecting more records and memories along the way. But suppose, for a moment, that some process we do not understand could build up in a person's mind all the memories relating to a life which that person had *not* lived, and could back up these memories with physical records, or at least with the impression of the existence of these physical records relating to that fictitious life. There is no way by which the person could detect the deception. He or she would believe that life to have been real, and would be led thereby to believe that throughout life, time had been *passing*, just as we do. We see that, through the agency of memory and record keeping, our belief in a *flowing* time can be explained, despite the fact that, as we have shown, such a conception is meaningless or nonsensical.

2.19 The possession of memory can explain also our belief in a fundamental difference between the past and the future. We *know* that past events have occurred because we can remember them, or because we have records of them. And if we have doubts we can often confirm our knowledge by comparing our memories or records with other people's. When allowance is made for human fallibility, the agreement between different people's recollections of events which they observed in the past is remarkable, and very difficult to explain if these events did not happen. Because some of these memories and records persist over long periods, it is clear that past events cannot be changed, for this would make false our memories of them. But the future, we believe, is another matter. We may sometimes make predictions, but this can be done only by surmise and calculation; predictions of future events do not have the same degree of certainty as memories of past ones. Indeed, we believe that *people* are sometimes able to affect future events in a way that they cannot influence the past. (The issue of *free will* is much too complex and controversial to discuss here; but many philosophers are now in agreement that the term "free will" becomes increasingly vacuous the closer one seems to a rational definition). So it is not surprising that we attribute the uncertainty of future events to the events themselves rather than to the limitations of our own faculty of memory, even though we now acknowledge that the past and the future cannot differ fundamentally in this way because the dividing moment which separates them, the "present" moment, has no objective existence.

2.20 We have discussed the Second Law of Thermodynamics, the (almost) universal tendency for systems to become increasingly uniform in temperature, to be in a state of greater *probability*, to present a greater degree of *dissipation*, as the value of t increases. Because this affects everything we experience, everything we think, everything we do, it is not surprising it becomes so familiar in the earliest days of our childhood that we accept it unthinkingly just as one of the laws of nature, like the effects of gravity, the impenetrability of solid bodies, or the heating effects of fire. We cannot be expected to know in these early days of our life that the fundamental laws of physics are in fact time symmetric, and that the asymmetry of most macroscopic phenomena is just an empirical fact pertaining in the particular universe we happen to inhabit. If great physicists such as Boltzmann could believe mistakenly that future events are dependent on past ones in an irreversible relationship, we need not be ashamed if we occasionally fall into the same trap. But we must try to minimise the number of such mistakes by constant vigilance.

2.21 Whenever we are in danger of being led astray by an unconscious belief that our own personal "now" actually has some objective or external existence, or that the space-time picture of the universe "changes" as time flows or progresses, there is a simple remedy. We should draw, or imagine, a representation of the four dimensional picture of the events we are considering, and firmly resist the temptation to draw on it the plane representing our "now", and even more firmly refuse to imagine such a plane moving along the time axis. As a simple illustration, the diagram represents a collision between two balls on a billiard table. Such a table is conveniently two-dimensional, so we do not need to discard a dimension in our representation. The illustration is static; the *movement* of the balls is completely represented by the *slope* of their world lines.

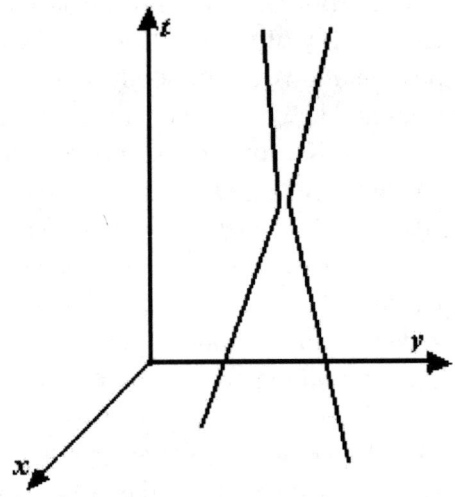

Fig. 2-4

- 36 -

2.22 The tendency to imagine things *moving* on a space-time diagram is more persuasive than may be imagined, and has led some capable thinkers into error. One sometimes reads about the possibility of things moving "backwards in time", and an author may illustrate the idea with a space-time diagram like that shown here. We are asked to believe that the line from A to B represents a particle moving forwards in time between two points, whereas if the movement on the diagram were from B to A, it would represent a particle moving in reversed time.

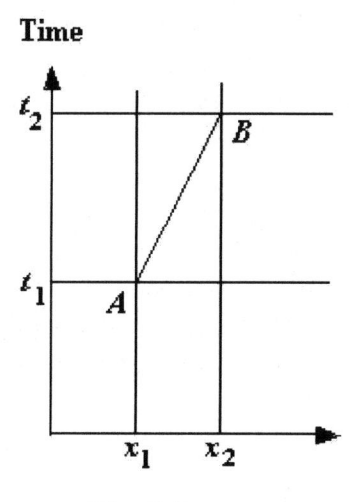

Fig. 2-5

But what is the difference between the behaviour of these two particles? The first is at x_1 at time t_1 and at x_2 at time t_2; the second is at x_2 at time t_2 and at x_1 at time t_1. So they are both at A at the same moment and both are at B at the same moment. They travel *together* from the point A to the point B, and may indeed be the same particle. This is clearly not how the authors expect us to think about it, for they had tried to distinguish a particle "moving" from A to B from one "moving" from B to A. We must always remember that the space-time diagram is static; it will serve its purpose of banishing our temporal illusions only if we firmly disallow any view of it which involves movement.

2.23 Much less rational is a lot of the literature relating to time travel into the past. Many a physicist or philosopher has been kept awake at night worrying about the "closed causal chains" which we shall be able to create when we master the art of time travel. The favourite story is of the man who journeys back to a time before his parents were born and murders his grandmother. Because of this, he could not have been born, so he couldn't have murdered his grandmother, and so he would have been born .... Critics point out that such a closed chain can not exist, since relativity teaches us that nothing can move faster than light. The world-line of any material body must always lie in a generally "time-like" direction on the space-time diagram, and with a suitable choice of scale no world-line can make an angle of more than 45° with the time axis. It is therefore impossible for it to double back on itself, and closed loops are impossible. But then topologists invent space-time topologies, such as cylinders or "worm-holes", wherein such closed loops *could* exist, and the heart-searching begins again. But what is the problem? If we look at the situation as laid out on a space-time diagram, with nothing allowed to change, even believers in time-travel must admit that the unfortunate grandmother either was murdered or she wasn't. If she was, her grandson never exists and the murder was committed by someone else.

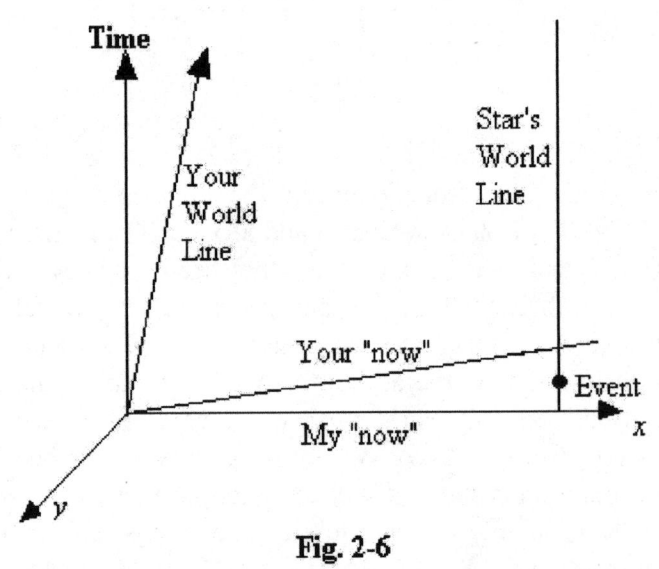

Fig. 2-6

2.24 A similar space-time diagram can be used to illustrate the distance- and time-transformations of Special Relativity, and this can drive home the point that the future and the past cannot differ in certainty in the way our intuition suggests. Suppose that I am standing on the platform at York while you pass on the London to Edinburgh express, and suppose that a large meteorite crashes onto a typical star 800 light years away. The diagram shows how our two "now" planes would be related, and shows it to be quite possible that the event occurred half an hour ago relative to your "now", but will not occur for another half hour for me. Does the event lie in the immutable past or in the uncertain future? It is clear that no such distinction can exist.

2.25 As we have seen, some systems and some processes are time-reversible and some are not. It is important for us to understand what sorts of process are of the one type, and which sorts are of the other, and here there is another powerful mental image we can call upon to help us in our task. We can imagine a process to be recorded in a video camera and played back in reverse; then if it is truly symmetric the reversed sequence of frames will still appear plausible. In fact we should not be able to distinguish the forwards from the reversed sequences, unless we know by some other means which actually occurred.

2.26 Suppose we have a pair of idealised billiard balls which are totally smooth and elastic, so that when they collide on a billiard table there is no loss of kinetic energy. Given the speed and direction with which each is travelling, and the orientation of the line of centres at the moment of collision, it is easy to calculate their speeds and directions after collision. And if we then imagine that these final velocities are reversed to produce another collision, the resulting speeds will be the reverse of what fomerly were the initial speeds. Being shown a video recording of such a collision, we would be unable to say whether it was being played correctly or in reversed time. But now suppose that one of these balls is replaced by a lump of clay, which does not fully recover its original shape after a collision; will the collision still be reversible? To find out, let us imagine the video is shown to us in reverse. A deformed lump of clay, bearing the imprint of the solid billiard ball, collides with such a ball, whereupon the imprint is removed and the clay becomes a perfect sphere. We see at once that this is impossible; the imprint is a result of the collision, and so must be present *afterwards*, not *beforehand*. This sort of process is so familiar that we assume it to be universal without giving it a second thought. If your leg is in collision with the table leg, it will be bruised *after* the collision, not before; the collision could not *remove* a bruise which was there beforehand. When we copy a picture in a photocopier, the new image is on the paper *after* it has been copied; the photocopier does not remove an image which was there beforehand.

2.27 And in this familiarity lies the danger. This was the error made by Boltzmann in his fallacious proof that the collisions of molecules in a gas must result in an increase of entropy, in other words must result in the energies of the molecules becoming *more* uniform rather than less. He made the assumption, which must have seemed very reasonable to him, as it still does to us, that the motion of a pair of molecules *after* they have collided is influenced by the fact that the collision did occur; their motions *before* the collision are not influenced by the fact that they will collide in the future. But view this assumption in reversed time and we see immediately that it introduces an asymmetry which is known not to be present; molecules are known to behave just like perfectly elastic billiard balls when they collide. We are more familiar with lumps of clay, bruised legs and photocopiers than we are with molecules of a gas, and we must ask why these latter behave so differently and counterintuitively. The essential difference is that lumps of clay are composite bodies; they have structure, and this structure is modified by a collision. Not only does the surface bear an imprint afterwards, but its molecules are set in motion which continues, resulting in a slight rise in temperature. Here is another example of what we have called "dissipation", and herein lies the time asymmetry, for the random motion of the clay molecules by virtue of their higher tempertaure is a more probable state of affairs than the equal velocities which they all possessed by virtue of the previous motion of the clay. And the condition of greater dissipation exists at a later time, i.e. at a greater distance from the highly ordered boundary conditions just after the big bang, than the preceding state of less dissipation.

2.28 As we study the strange world of the quantum, we must frequently view processes as static models in the four dimensions of space-time. And we must ask ourselves how the explanations we formulate would appear in reversed time, to prevent ourselves being misled by the deeply ingrained false impressions our intuition gives of the nature of time.

Chapter 3: Probability and Quantum Amplitudes

3.01 Many events in everyday life appear to be random, and the only way we have of predicting them is to use the methods of probability. There is no way by which we can know in advance the result of tossing a coin, rolling a die or drawing a raffle ticket from a hat, but knowledge of the probabilities involved is often useful. In fact none of the everyday situations in which we use the methods of probability are really random, for all are examples of deterministic processes whose outcome could be calculated if we had sufficient data and sufficient mathematical skill. If we know the exact position where a tossed coin is released and the exact value of its linear and angular momentum, then we have sufficient information to work out whether it will fall head or tail uppermost. But the necessary precision of the data, and the complexity of the calculation, remove any possibility that we could do so in practice. The factors which we would need to take into account are "hidden variables" which we could never determine with sufficient accuracy.

3.02 Probability plays an even bigger part in Quantum Mechanics, for almost all events and processes seem to involve an element of randomness. When conducting an interference experiment we can have no idea where a particular particle will strike the screen, but can often specify accurately the probabilities of its arriving within different specified areas. When a photon approaches a pair of polarising filters we cannot know in advance whether it will be transmitted or absorbed, but when a large number of photons are involved we can calculate with precision the proportion which will be transmitted, and hence the probability that an individual photon will be. We have no idea when a radioactive nucleus will disintegrate, but we know accurately the probability that such a nucleus will do so in any given time interval, from which we can calculate the half-life of a sample containing a large number of atoms, and the proportion remaining after any required time.

3.03 The question of whether quantum phenomena also exhibit apparent randomness only because they are determined by hidden variables of which at present we have no understanding, or whether they are truly random in a way that macroscopic systems never are, was vigorously debated during the second quarter of the twentieth century. In particular the famous discussion between Niels Bohr, who was the chief spokesman for the most widely held views on the interpretation of quantum phenomena, and Albert Einstein, who was convinced that quantum behaviour would indeed prove to be deterministic when we discovered the hidden variables which dictated its apparent randomness, provided a fascinating example of great brains wrestling with a difficult problem. Although the argument was not resolved during the lifetime of either man, the depth of thinking which it evoked on both sides did much to clarify our ideas. Today, in the light of experimental developments which have occurred since Einstein's death in 1955, it seems likely that he was on the losing side. Einstein refused to accept two of the most strange aspects of Bohr's viewpoint, the indeterminacy of quantum behaviour, and the apparent transmission of influences at speeds greater than that of light. This latter principle, which has come to be known as "non-locality", seemed at the time to violate the dictates of Special Relativity. But during the last quarter of the twentieth century many experiments appear to have shown that we cannot escape from non-local influences even if we accept that their randomness is a result of hidden variables. It is likely that Einstein would have admitted defeat in view of this more recent evidence, for it seems impossible to describe the quantum world wholly in terms of classical principles, as he had hoped. But whether or not quantum phenomena are really random, we must still use the methods of probability theory to describe them, just as we do for macroscopic behaviour which is not truly random, such as that of coins and dice.

3.04 The concept of probability is much more difficult to define than most people imagine, and no branch of mathematics provides more fertile ground for misunderstanding and paradox. Sometimes the results are amusing, as when we seem to find several different answers, all plausible, to the same problem. Sometimes they are frustrating, and throughout the twentieth century the interpretation of Quantum Physics has been hampered by misunderstandings and lack of clarity in the application of probability theory. But sometimes they can be disastrous, as when they give rise to miscarriages of justice which commit innocent people to long prison sentences. We shall illustrate all these types of mistake in a later section of this chapter. In order to prevent ourselves making similar mistakes it is necessary to study the concept of probability in some detail.

3.05 We use the techniques of probability theory only in relation to propositions or events about which we have incomplete knowledge, usually because they lie in the future. If we knew exact details of everything that has ever happened or ever will happen, probability would no longer be necessary or meaningful. It follows that any value we quote for the probability of an event is relative to some corpus of knowledge that we have about that event, or that we choose to hypothesise about it, and will change if that knowledge changes. It is often supposed that an event can have an intrinsic probability, a property possesed by that event, just like the time or place of its occurrence, but this belief shows a fundamental misunderstanding. Probability is merely a relationship between an event and one person's knowledge of that event. The value of a probability may well differ for different people, and may differ for one person at different times, as that person gains additional knowledge of the event. This answers the rather silly question of whether the odds of a coin landing "heads" remains 1:1 after we have actually seen how it landed. If we choose to use the additional knowledge we have after looking at it, then the probability of a head becomes 0 or 1, as the case may be. If we do not, then it remains 1/2.

3.06 Attempted definitions of probability may be described either as "subjective", if they refer only to a particular person's beliefs regarding unkown or future events, or "objective", if framed in terms of the actual occurrences of such events themselves. Mathematicians seem to prefer objective definitions, while philosophers are more likely to subscribe to the subjective. The philosopher often thinks of the probability of event A as his "degree of rational belief that A will occur", an interpretation particularly associated with Maynard Keynes (1883-1946) and Rudolf Carnap (1891-1970). Thus he may claim to be "one half certain" that the coin will land heads, or 1/52 certain that he will draw the Queen of Hearts. The writer cannot accept this as a definition. Surely we cannot have "degrees of belief". There are some propositions which I believe to be true, and some that I believe to be false, but if I am convinced neither one way nor the other about a proposition then I have no belief concerning it. I cannot be certain that the next throw of the coin will result in a head, and so I do not believe that it will, and nor do I believe that it will result in a tail; maintaining that one can have "half a belief" is an attempt to divide the indivisible. Those who claim to have "degrees of rational belief" are using the word "belief" in a sense that I do not understand. The phrase sounds plausible at first, but those who use it really mean something different, probably something nearer to the mathematical definition of probability, but which they cannot put into words because of unfamiliarity with mathematical notation and convention.

3.07 One attempt to overcome this kind of objection is provided by the "principle of indifference". This proposes that if we have no reason to believe one proposition rather than another, we should assign them equal probabilities. It is easily seen, however, that this is untenable because of the inconsistencies it entails. Suppose you pick three names at random from a telephone directory, Mr. A, Mr. B and Mr. C. You have no reason to know whether A is taller or shorter than B, so by the principle of indifference you assign equal probabilities to the two propositions. In other words, the probability that A is taller than B is one half. By a similar argument, the probability that A is taller than both B and C is also one half. It follows that whenever A is taller than C, then A is also taller than B. This is absurd, and shows that our original premise, the principle of indifference, is false.

3.08 Another approach is the "Personalism" of Bruno de Finetti (1906-1985) and Frank Ramsey (1903-1930), defining the probability a person attributes to a certain event in terms of the betting odds he would wager on it. Again this is too vague to furnish a reliable definition, for people differ widely in their readiness to take risks. Some young men will stake their lives on the belief that they can control their cars at speeds which are clearly too high, while some older people will not venture out of doors for fear of being attacked. This is no basis for assessing the probability of being involved in road accidents or becoming victims of crime.

3.09 In fact it seems that no truly subjective definition of probability is viable, for such a definition must be derived at some level from an objective one. It is mistaken to think we know *a priori* that a coin lands heads with probability 1/2. We have already had some experience, however remote, of probability defined mathematically, and hence objectively. We claim to be one-half certain the coin will land heads only because we can visualise it being tossed repeatedly, and know that the number of heads is likely to be about one half the number of throws, or alternatively because we understand that one half of the coin's possible starting conditions lead to a head.

3.10 If we are thus driven to define probability mathematically rather than subjectively, why should this prove difficult? There are many situations, in gambling and elsewhere, when the probability of some event is obvious, with no room for disagreement. As an example, suppose you are asked for the probability of throwing an *even* number when a die is rolled. You know there are six possibilities, of which three are even, so the probability must be 3/6, or 1/2. So let us ask what are the distinguishing features of situations in which probabilities are fairly obvious. Firstly, they must be repeatable. We can roll a die as often as we like, and this is what we would do if our assertion of the above probability were challenged. When a possible situation is totally unique, and cannot be repeated, it is meaningless to ascribe to it a probability, for how could we ever confirm that the value we give it is correct? Secondly, situations governed by probability always involve an event (in this case the throwing of an even number) which, in different trials, sometimes happens and sometimes does not. In many such situations, the proportion of trials in which the event happens, while it can take any value between 0 and 1, is found to concentrate around some particular value as the number of repetitions increases. This is the value we have in mind when the probability of an event seems obvious. Thirdly, while each trial may result in any one of a number of outcomes (in this case six), the event itself is related to a particular subset of these outcomes, in this case 2, 4 or 6.

3.11 We can illustrate these facts on a diagram. The initial conditions, or the given situation, may be represented as the values of n variables, x_1 to x_n. We suppose that the first m of these (e.g. the size and shape of the die) are held constant during all the trials, while the others (e.g. the position and speed of the throw) are allowed to vary unpredictably. Thus x_{m+1} to x_n are the hidden variables, and if we deny the existence of these in quantum situations, all the variables are fixed. We represent the outcomes of the trials by y_1, y_2, ... , and the set of all possible outcomes (in this case 1, 2, 3, 4, 5 or 6) we call collectively the "sample space". The event E (e.g. the throwing of an even number) comprises a particular subset of the outcomes (2, 4 or 6).

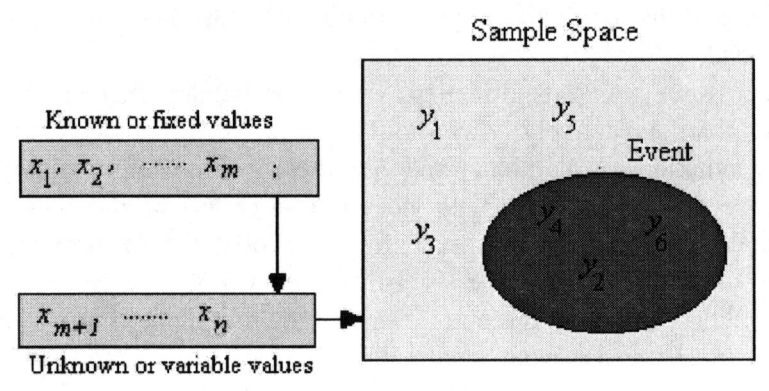

Fig. 3-1

Our first attempt at an objective definition relies on the fact that the possible outcomes of an experiment, which we have represented by y_1, y_2 ..., will often be equally likely. The six possible results when a die is rolled satisfy this condition, provided the die is not loaded. Then if there are s equally likely ways in which an event can occur, and r of these are "successful", the probability of success is r/s. In such cases this would seem to provide an unambiguous meaning for "the probability of the event occurring". The trouble is that the definition is useless because it is circular; "equally likely" is just another way of saying "equally probable". Our definition of probability is meaningful only if we know already what probability means.

3.12 A second approach to the problem is to use a method known as "range theory". The result of a trial is often uniquely determined by the values of the variables $m+1$ to n in the diagram above, and we suppose these to vary randomly. Can we not define the probability of "success" by expressing the range of possibilities of these $n-m$ variables under which success is achieved as a proportion of the total range of possibilities? In other words, can we not take the average value, with success counting as 1 and failure as 0, averaged over these random variables? Unfortunately no; not with any certainty that the method will yield a unique value. It has long been known that this range method can lead to conflicting values for the same probability, and a good example of this was described by J. Bertrand (1822-1900) in 1889.

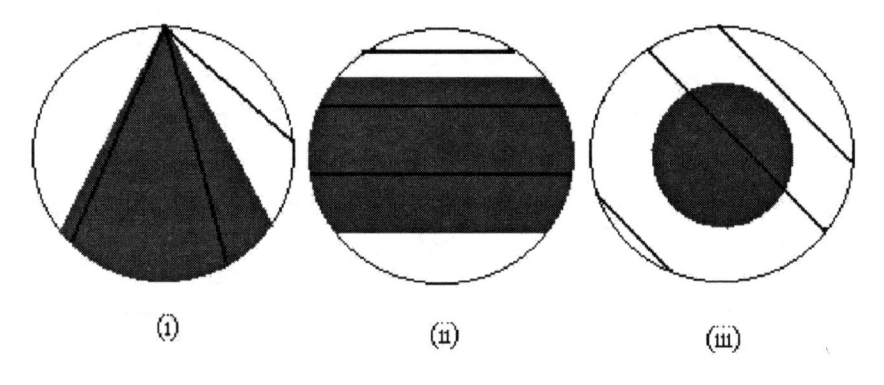

Fig. 3-2

3.13 He asked what is the probability that a chord drawn at random in a circle is longer than the side of an equilateral triangle inscribed in the same circle. If we consider all chords drawn through a particular point on the circumference (see (i)), the condition is satisfied only by those chords which make an angle of less than 30 degrees with the radius through that point, out of a total possible range of 90 degrees, so the required probability must be 1/3. But if instead we consider all the chords drawn perpendicular to a given diameter (see (ii)), the condition is satisfied whenever the centre of a chord is less than half a radius from the centre, so the required probability must be 1/2. Or thirdly, we can see that the condition requires the centre of the chord to lie within a circle of radius one-half that of the given circle, i.e. within a circle of area one quarter of the given circle (see (iii)). So the probability must be 1/4. A definition which gives three different answers to the same question is clearly unacceptable. Despite the fact that many writers believe range theory to provide a sound definition, and a means of determinig probability values in practice, we must reject it for either purpose because of its unreliability.

3.14 Both of the above methods of assigning probabilities may be described as *a priori*, for they depend upon knowledge we have of a system before conducting any trials. A third method, the empirical method, defines probability in terms of the actual frequencies observed when a practical experiment is performed many times, rather than by consideration of causes. This approach is associated with John Venn (1834-1923), but this too is not without difficulties. Let s denote the number of trials, and r the number of "successes", and define probability as the observed value of r/s. This is easily seen to be unsatisfactory, for it will give a different value each time the experiment is repeated, and will seldom give the value we know intuitively it should. If a coin is tossed 1000 times it is unlikely the number of heads will be exactly 500: it is almost equally likely to be 510 or 520, giving values of 0.51 or 0.52 for the probability which we know ought really to be 0.50. We might claim that we can always get as close as we wish to the correct value by repeating the experiment a sufficient number of times, but this also is false. If the coin is tossed a million times it is true that 510000 heads are unlikely, but they are not impossible. Indeed if the million tosses are themselves repeated a sufficient number of times, we can calculate when it will become more likely than not that at least one of the attempts will yield 510000 heads. We cannot be sure of getting as close as we wish to the correct answer by any pre-specified number of trials, however great.

3.15 We can attempt to put the above method on a firmer mathematical footing by defining the probability not as the value of r/s, but as the limit of r/s as s tends to infinity. But this fails equally, for it can be shown that if the sequence of successes and failures is truly random, then r/s does not, in fact, tend to a limit. The strict meaning of "r/s tends to the limit p as s tends to infinity" is that, given any small number q, a value s_o for s can be found such that r/s differs from p by less than q for all values of $s > s_o$. Clearly this will never be true; it is always possible for a run of successes to occur long enough to raise the value of r/s, or a run of failures to lower the value of r/s, so that it lies outside the limit prescribed. Nonetheless, if we ignore this difficulty, and try to assess a limiting value by observing a long series of trials, we shall usually obtain a result close to the correct one. The error could, in fact, be of any magnitude, but with a sufficiently long sequence, it will be substantial only very rarely.

3.16 The previous section has provided three possible definitions of probability. Each is imperfect in some respect, as we have shown, but each does sometimes give a useful method of finding probability values in practical investigations, with little risk of error.

3.17 The occurrence or non-occurrence of an event E depends upon which of the y alternatives occurs, as our diagram shows. Often we have good reason to believe that these alternatives are all equally likely (while realising we cannot put this belief on a firm logical footing), and then we know the probability of E will be the proportion of these alternatives which entail E. The symmetry of a die or a coin is sufficient reason to allow us to believe the outcomes to be equally likely when we roll a die or toss a coin. The probabilities of various combinations of result are then easy to calculate.

3.18 In reaching this conclusion we are unthinkingly making use of range theory. The result of spinning a coin obviously depends upon the speed with which it is spun, and because we cannot accurately control this speed we assume an equal likelihood for heads and tails. This is reasonable because we know a small difference of speed can cause a large difference in the number of times the coin turns over while in flight.

3.19 But there are often cases in which an analysis of causes gives little guide to the probabilities we are attempting to find, and in such cases we must rely on the empirical method. We conduct a large number of trials, and assume the probability to equal the proportion of outcomes which are "successful". Thus, to find the probability that a single photon arrives within a particular region of the screen in an interference experiment, we observe a large number of photons, all prepared in the same way, and we note the proportion which arrive in this region. Notice that, if we reject the "hidden variable" theory, we can indeed prepare our photons all in the same state, in contrast with what happens in corresponding macroscopic experiments. When tossing a large number of coins we cannot ensure that they all have the same starting conditions; the variables x_{m+1} to x_n are outside our control. But in the quantum case these values do not exist; if we reject hidden variables, then $m = n$, and *all* the initial conditions can be determined by the experimenter.

3.20 It is necessary to emphasise again that the value of a probability assigned by a person to a particular event depends on that person's imperfect knowledge of the event. If you know that the four Aces have been removed from a pack of cards, for you the probability of drawing a Queen is 4/48, or 1/12, while for me it will remain 1/13. Neither of these values should be regarded as wrong; each is a correct evaluation based upon one person's knowledge of the situation. A simple error known as the "gambler's fallacy" illustrates the importance of specifying carefully what knowledge we are taking into account. If a coin has been observed to fall heads on ten successive occasions, many people believe wrongly that the likelihood of a head on the eleventh throw is then less than 1/2, because of what they call the "law of averages". Such people are confusing two questions. If a coin is tossed eleven times and we know none of the results, the probability of all eleven being heads is less than 1 in 2000. But if we know that the first ten results are indeed heads, then the probability of the eleventh also being a head is just 1/2 as always.

3.21 Not only the ignorant can make such errors. The renowned physicist Paul Davies describes a simple experiment in which a stream of electrons builds up an interference pattern on a screen, and writes,

> These results are so astonishing that it is hard to digest their significance. ... How does any individual electron know what the other electrons ... are going to do? (*Other Worlds*, Penguin, 1980)

The results are indeed surprising, but not for this reason. Does Davies ask how a coin knows which way it will fall on subsequent occasions? It is because the coin does *not* know, because its results are independent of each other, that the characteristic distribution of results is built up.

3.22 The knowledge we take into account may be just hypothetical; even if I know the pack to be complete I can still ask the question, "If the four Aces *were* missing what would be the probability?" In this type of question we are making an adjustment to the first set of x values in the diagram above, the "known or fixed" values. But it is important not to allow this to happen inadvertently; such errors account for many of the well-known paradoxes associated with probability theory.

3.23 A dangerous error arises when we fail to specify sufficiently clearly the event E whose probability we want. A good illustration is provided if we toss three coins, and ask what is the probability that they all fall showing the same face, i.e. all heads or all tails. There are clearly eight equally likely results, HHH, HHT, HTH, etc., and two of these satisfy our condition, namely HHH and TTT. The required probability is therefore 2/8 or 1/4. But what is wrong with the following argument? When we look at three coins, there must be two showing the same face. The probability that the third coin agrees with them is then 1/2, not 1/4 as we obtained previously. The trouble here is that we are not specifying E clearly; we do not know which two coins will be the same, nor whether they will be HH or TT, and so the requirement we are imposing on the third coin is not fixed either. While enquiring whether the third coin's y value lies within the event E we are shifting the goalposts through which we require it to pass.

3.24 Many paradoxes arise from carelessness in specifying the sample space within which a probability is to be determined. If Mr. Smith tells us he has two children, of whom the elder is a boy, we would rightly suppose the probability of the younger also being a boy to be 1/2. But if he tells us he has two children, *one* of whom is a boy (without saying which), the probability of the other being a boy is now 1/3, for our knowledge is different. The new sample space contains three equally likely possibilities, namely BB, BG and GB (in each case specifying the elder child first), only one of which meets the required criterion.

3.25 A serious mistake sometimes occurs in a court of law. Let us suppose a man living in England is accused of a murder or a rape. The only evidence against him is a sample of bodily material left at the scene of the crime. If the forensic scientist tells the jury that this sample could have been left by only one person in a million, it is quite possible the defendant will be convicted on this evidence, for the jury are likely to conclude that the probability of him being innocent is only one in a million. In fact it is nothing of the sort. If we suppose a total adult male population of twenty million, then we can expect about twenty men to match the forensic sample. Only one of these men is guilty and the other nineteen are innocent, so the probability of the suspect being innocent, *if no other evidence is available*, is 19/20. The jury are not being asked to decide whether the suspect matches the sample; this piece of information is given. They are required to decide if he, among those who do match, is the one responsible for the crime. They mistook both the sample space and the subset of outcomes required.

3.26 The *New Scientist* journal (13 December 1997) quotes several cases in which a suspect has been found guilty on this sort of evidence, with neither the prosecution nor the defence noticing the fallacy. In one of these cases the accused had to serve a seven year jail sentence after the failure of two appeals.

3.27 Many people try to discuss the probability of hypothetical events which by their nature are unique, despite the fact that such a probability is meaningless. The actuaries employed by insurance companies work constantly with probabilities in assessing premiums. In most cases these probabilities will be determined by empirical methods; life insurance premiums are based on past statistics of life expectancy, and insurance against motor accidents or wet days will depend on past accident rates and weather records. But occasionally clients may seek cover for events which are essentially singular, in that no sequence of previous trials can exist. In these cases insurers must rely upon intuition. What would you expect to pay to cover the cost of rebuilding your London home if it is struck by a meteorite during the next fifty years, or replacing your workforce if it is captured by aliens from another planet? We never ought to talk about the probability of such events.

3.28 Our final example of a paradox is taken from Quantum Mechanics, and has engendered more than sixty years of argument. It concerns the "EPR" experiment devised by Einstein and two colleagues 1935, which seems to lead to impossible results whenever it is performed. While we admit that the results are unexpected, we maintain that the apparent contradiction arises from applying probability methods without defining properly the sample space involved. A full treatment will be found in a later chapter.

3.29 We have to use the methods of probability theory when solving problems at the atomic level because quantum systems are not deterministic. By this we mean that the present state of a system does not contain sufficient information to enable its future state to be calculated unambiguously. Experiments in particle physics are often conducted by preparing large numbers of identical particles in identical states. All the variables of which we are aware, such as position, momentum and spin, are set to identical values, so far as possible, and yet the particles do not all behave in the same manner, and the only way of relating their future performance to their present state is by means of probabilities.

3.30 The same applies to macroscopic systems, such as coins, dice or roulette wheels, which do behave deterministically, but whose behaviour is so complex that we cannot possibly know the values of all the relevant variables. The behaviour of elementary particles, however, displays an additional complication for which there is no parallel in the world of familiar experience. Ordinary probabilities are expressed as single numbers lying between 0 and 1, with 0 representing impossibility and 1 certainty. But many examples of quantum probabilities can be described only by using *two* numbers, or alternatively a *two-dimensional vector*. The best way to visualise such a vector is as a small arrow, such as one would use to represent a *velocity* or a *force*. The two numbers would then stand for the *length* of the arrow, taken to represent its strength or magnitude, and the *angle* it makes with some fixed direction such as the x-axis. The reason for this complication, and that an ordinary probability value will not suffice, lies in the fact that a quantum probability possesses, in a way which we cannot understand, a *phase angle* as well as a magnitude, and consideration of these phase angles plays an important part in the calculation of quantum effects, particularly in phenomena involving interference. So in describing a quantum probability we must quote both the magnitude and phase of such a vector. However, mathematicians often use an alternative description, and give us instead the x and y co-ordinates of the *tip* of the arrow, assumed to be drawn from $(0,0)$ to (x,y), and they then go a step further and use just a single *complex number*, of the form $z = x + iy$, where i is the square root of -1. Non-mathematicians must believe us when we tell them that such a procedure does considerably simplify calculations involving these *quantum amplitudes*, or *probability amplitudes*, as they are called. (It is unfortunate that the *y* component of such quantities has come to be called the "imaginary" part. In fact it is just as real as the *x* component, and the vector represented by such a complex number can be just as real as the force you exert when dragging a piece of furniture across the floor. The only strange fact in the application of complex numbers to quantum mechanics is the need for *two* numbers rather than one.) When it is required to convert a quantum amplitude to an ordinary probability, we find that it can be done by *squaring* the length of our arrow, or the *magnitude* of the complex number. We write this as $|z|^2$, and the value is given by the formula $x^2 + y^2$.

3.31 We often need to multiply probabilities together. If A and B are independent events, then the probability of the combined event "A *and* B" is found by multiplying together the probability of A and the probability of B. The probability of drawing a red card from a pack is 1/2, and the probability of drawing an ace is 1/13, and so the probability of a red ace is 1/2 multiplied by 1/13, which comes to 1/26. It can easily be seen that the same rule applies to quantum amplitudes. We can find the amplitude for a compound event, when we know that for each separate event, by multiplying together the two amplitudes, and then if we want the *probability* of the compound event, we just square its magnitude in the usual way. In fact it does not matter whether we multiply the magnitudes and then square the amplitude, or square the two separate amplitudes and then multiply, for it can be shown that $|z_1 \times z_2|^2 = |z_1|^2 \times |z_2|^2$.

3.32 And we sometimes must add probabilities. If A and B are mutually exclusive events, then the probability of the event "A or B" is found by adding the probabilities of A and of B. The probability of drawing a spade is 1/4; so the probability of drawing a red card or a spade is 1/2 + 1/4, or 3/4. But we must be careful when dealing with the quantum amplitudes of two such events, for it is *not* true that $|z_1 + z_2|^2 = |z_1|^2 + |z_2|^2$. In fact the corresponding rule for quantum amplitudes is much stranger. If we have set up an experiment in which we can determine not only that "A or B" occurs, but *which* of the two alternatives, then the same rule applies as with normal probabilities; we must find the separate probabilities by squaring magnitudes, and the final composite probability by adding these in the usual way. But if we know only that "A or B" occurs, and it is inherently impossible to know which alternative, we must add the complex amplitudes first, and *then* square the magnitude, or we get the wrong answer. This situation occurs, of course, in interference experiments, where particles have two or more possible routes to their destination, and where we cannot find out which route has been followed without destroying the interference effect, and to calculate probabilities we must add together the relevant probability amplitudes *first* and then find the squared magnitude. If instead we square the magnitudes separately and then add them together, we shall thereby lose track of the phase angles which determine the interference effects. This strange fact lies at the heart of the most perplexing quantum paradoxes, and the reader is urged to be certain it is fully understood before reading the remainder of the book.

Chapter 4: The Problems

4.01 Quantum theory was born in 1900. Towards the end of the previous century many of those engaged in physical research had believed their work to be almost over. The world contained only matter and radiation; all matter consisted of electrons and protons, and all radiation of electromagnetic vibrations in the ether. True, there were a few outstanding unsolved problems concerning radiation, but these should soon be overcome. Little did the physicists of the day guess that within a few years fresh problems would present themselves which were so intractable that they would still remain unresolved a century later.

4.02 The first question to be addressed in the new century concerned the radiation from hot bodies such as stars. The distribution of energy from such bodies and the way it depended on the temperature were well understood, but several attempts to derive these relationships by calculation failed. The failure was not just in the accuracy of the results; the answers obtained were nonsensical, showing that something fundamental was wrong in our understanding of the process of radiation. In 1900 Max Planck suggested a way out of the difficulty. He proposed that light was always emitted in small packets or *quanta*, the amount of energy in each quantum being proportional to the frequency of the light wave it contained. Symbolically, $E = hf$, where h is the constant number mentioned in Chapter 1. Calculation showed that this hypothesis led to exactly the right energy-curve, and the correct relationship with temperature. Planck's constant, h, could be calculated accurately, and was found to be a very small quantity, which explained why the effects of this quantised radiation of energy had not been suspected earlier.

4.03 This new theory became firmly established as it was found to explain with equal success some of the other problems associated with radiation. In 1905 Einstein applied it to the emission of electrons from the surface of a metal when light fell upon it, the *photoelectric* effect. The manner in which their energy depended on the intensity and wavelength of the light could not be explained by the old wave theory, and Einstein showed that quantum theory provided a full explanation. A few years later, Arthur H. Compton performed some remarkable experiments in which a block of graphite was illuminated by x-rays, and the wavelength of the scattered radiation was measured. Using Planck's energy formula, Compton showed that the energy of the scattered quanta depended upon the direction in which they moved just as if they had bounced off the atoms of graphite like billiard balls from the cushion of a billiard table. Here the x-ray quanta, or *photons* as they had been named, were behaving not like packets of radiation but like material particles.

4.04 Then in 1924, Louis de Broglie suggested that this dual nature of light, behaving sometimes like a wave and sometimes like particles, might apply also to material particles such as the electron. In fact he proposed that every particle of matter was associated with a wave when it moved. If v stands for the speed of the particle, and m its mass, the wavelength of the wave was given by the formula $\lambda = h/mv$. This hypothesis was soon confirmed when diffraction and interference experiments were performed on electrons. Solid bodies behave, just as do light rays, sometimes as particles with precise positions and velocities, and sometimes as waves spread over a region of space, and having measurable wavelengths.

4.05 Two years later Edwin Schrodinger devised the equation which bears his name, and which describes the motion through space of the wave associated with any particle whose momentum and energy are known. At first Schrodinger believed that, when a particle needed to be described as a wave, it had become spread out throughout a volume of space, and the wave function which his equation gives us, and which he represented by the Greek letter Ψ, provided a measure at each point of the *density* of the distributed particle there. But it was soon realised that a much more satisfactory interpretation was obtained if Ψ was regarded as a measure of the *probability density* that the particle would be found at that point. The particle itself had not become dispersed, and the Ψ value at any point tells us just how likely we are to find it there. In fact Ψ is often a complex number, as we described in Chapter 3, and the associated probability density is given by the square of its magnitude, $|\Psi|^2$.

4.06 As understanding of sub-atomic physics improved, techniques were devised for making accurate measurements on particles and waves. But it was found impossible to devise experiments which would measure simultaneously, with a high degree of accuracy, both the position and the velocity of a particle. The greater the precision of one result, the more inaccurate was the other. Then in 1927 Werner Heisenberg discovered the *uncertainty principle* which is named after him. He showed that these inaccuracies were not due to poor experimental techniques, but that they were an inevitable result of the properties of waves which, in other contexts, had been known for many years. Suppose we know the position in the x-direction of a moving wave to a certain degree of accuracy, which we shall denote by dx. This means that the wave-packet has a length of dx, and contains only a limited number of complete waves, as shown in the diagram. Now it is known from general wave theory that such a packet, because of its limited extent, must comprise not just one wavelength, but a range of wavelengths which, by their mutual interference, can limit the packet to a finite size.

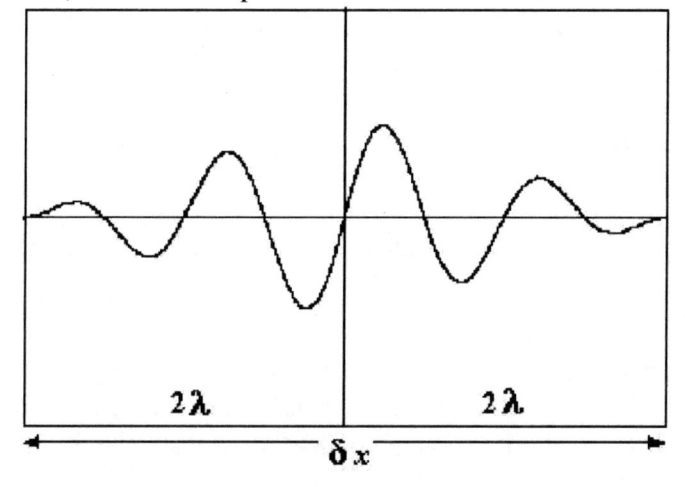

Fig. 4-1

The fewer the number of wave crests within the packet, the larger the range of wavelengths which it must contain. The packet illustrated extends over about four wavelengths, and so it must be made up from a range of pure sine waves whose minimum and maximum wavelengths differ from the average by about one-quarter of this average. We know also the relationship between the de Broglie wavelength of a particle and its velocity, $\lambda = h/mv$, and so the spread of wavelengths is a measure of our uncertainty of the particle's velocity. If we let p stand for the particle's *momentum*, which equals mv, we have $\lambda = h/p$, and it is easy to show that, for the wave packet illustrated, the uncertainty of position, dx, and the uncertainty in momentum, dp, multiply together to give h. In general, any simultaneous measurements of position and momentum of a particle result in $dx.dp$ being at least equal to h, and this is Heisenberg's principle.

4.07 Great progress was made in quantum physics during the first quarter of the twentieth century, but the problems and the perplexities were accumulating. How can an electron, say, be at the same time a particle, with a definite position and negligible size, and a wave, extending over a wide region of space? And how can we account for the fact that, if we wish to visualise the behaviour of a quantum system without actually observing or measuring it, we must follow the development of the wave, but whenever we make a measurement the wave appears to *collapse*, and a new wave is required to describe its future development? And what is the nature of the *uncertainty* which Heisenberg tells us is inherent in the system, and not just the result of poor experimental technique? Is the world not *deterministic* at the quantum level? Is it no longer true that identically prepared experiments lead to identical results, as they do in the world of common experience? Perhaps quantum effects really are deterministic, but we have not yet discovered the "hidden variables" which control them, in the same way that the hidden variables of standard mechanical theory control the motion of a spinning coin.

4.08 Early in the development of quantum ideas Einstein and Bohr recognised very clearly the dilemma posed by the dual particle/wave nature of quantum objects, displayed forcefully by experiments where *interference* effects occur, effects that can be attributed only to waves. Bohr wrote as follows:

> The [problem] is strikingly illustrated by the following example to

which Einstein very early called attention and often reverted. If a semi-reflecting mirror is placed in the way of a photon, leaving two possibilities for its direction of propagation, the photon may either be recorded on one, and only one, of two photographic plates situated at great distances in the two directions in question, or else we may, by replacing the plates by mirrors, observe effects exhibiting an interference between the two reflected wave-trains. In any attempt of a pictorial representation of the behaviour of the photon we would, thus, meet with the difficulty: to be obliged to say, on the one hand, that the photon always chooses one of the two ways and, on the other hand, that it behaves as if it had passed both ways. (*Discussions with Einstein*, 1949)

In the latter case, the production of interference fringes by using two mirrors, the photon always arrives at the screen as a single particle, and yet interference is displayed by the differing probabilities of its arriving at different parts of the pattern. For example it never reaches the centre of one of the dark fringes, and this shows that the wave must have been reflected by both mirrors. Which way has the actual photon itself gone? The depth of the mystery increases if we try to detect the photon while it is traversing the apparatus, for any successful attempt, whatever means we employ, shows that it has gone via *one* mirror only, but at the same time *destroys* the interference pattern. Nature will not reveal how a particle can perform an interference trick.

4.09 The "measurement problem" is really just another manifestation of this wave/particle dilemma. Because of the possibility that interference effects may occur, we must trace the behaviour of a quantum system between one "measurement" (or "observation") and the next by means of its *wave* representation, often making use of Schrodinger's wave equation. But the next observation seems to make the wave-function collapse, and the results of the observation always involve the new positions of *particles* rather than a new form for the wave. Unless and until another observation is made, no collapse occurs, and this has led some theorists to maintain that the very act of "looking at" a quantum system in some way determines its future development, as was described in Chapter 1.

4.10 The Copenhagen interpretation is probably the most

commonly held view on how these difficulties can be resolved. Bohr acknowledged that the particle and the wave were incompatible pictures of the quantum world, but side-stepped the issue with his notion of "complementarity". He pointed out that our knowledge of the micro world could be obtained only through experiments using macroscopic apparatus, and that any experiment asking questions about particles could not be used for waves, and *vice versa*, so at no time is a contradiction exposed by any one experiment. Thus an attempt to measure accurately the *position* of an electron (a "particle" experiment) rules out any prospect of measuring at the same time its *momentum* (which must involve a knowledge of its wavelength, requiring a "wave" experiment). Bohr writes about -

> the impossibility of any sharp separation between the behaviour of atomic objects and the interaction with the measuring instruments which serve to define the conditions under which the phenomena appear. ... Consequently, evidence obtained under different experimental conditions cannot be comprehended within a single picture, but must be regarded as complementary in the sense that only the totality of the phenomena exhausts the possible information about the objects. Under these circumstances an essential element of ambiguity is involved in ascribing conventional physical attributes to atomic objects, as is at once evident in the dilemma regarding the corpuscular and wave properties of electrons and photons. (*ibid.*)

Bohr rejected the idea of "hidden variables" on the grounds that, if they were essentially unobservable, they could play no part in a theoretical interpretation. He regarded the apparent indeterminacy of quantum phenomena as a fundamental component of their nature. And he maintained that the "collapse of the wave function", when an observation was performed on a micro-event, occurred at the point in the causal chain between the event and the observer where an "irreversible amplification of the effect" had taken place.

4.11 An alternative viewpoint, associated particularly with Eugene Wigner, regards this point of irreversibility as too poorly defined to identify something as clear-cut as the waveform collapse. His contention is that the waveform continues to provide the only possible description of what is happening right up to the point at which knowledge of an event enters a *conscious mind*, which he claims is an occasion of much greater certainty than an "irreversible amplification". A more extreme attitude to this question --

> ... has led Wigner and John Wheeler to consider the possibility that, because of the infinite regression of cause and effect, the whole universe may only owe its 'real' existence to the fact that it is observed by intelligent beings." (*In Search of Schrodinger's Cat*, John Gribbin, 1984)

4.12 A number of physicists do not believe that this intervention of conscious minds is an essential requirement for the universe to be "real", but subscribe instead to the "many universes" doctrine. David Deutsch describes it thus:

> The idea is that there are parallel entire universes which include all the galaxies, stars and planets, all existing at the same time, and in a certain sense in the same space. And normally not communicating with each other. But if there were no communication at all there wouldn't be any point to our postulating the other universes. The reason why we have to postulate them is that, in experiments on a microscopic level in quantum theory, they do in fact have some influence on each other. (*The Ghost in the Atom*, ed. Davies, 1986)

The purpose of this theory is to circumvent the problem of the collapse of the wave function. Whenever a quantum system must choose between two states the *whole universe* splits into two, identical in all respects except that in one the first choice has been followed, and in the other the second. Each universe contains a separate copy of every conscious being, including you and me, complete with all our memories of the past, and each copy of us continues to live a separate diverging life, unaware of the existence of the other. In one version of the theory the number of universes continually increases as branching occurs, but in another version, all the universes already exist as indentical copies of each other, and a branching consists just of identical worlds becoming differentiated as a result of one quantum choice. Deustch subscribes to this view.

> In my favourite way of looking at this, there is an infinite number of them and this number is constant; that is, there are always the same number of universes. ... When the choice is made, they partition

themselves into groups, and in one group one outcome happens and in the other group another outcome happens. (*ibid.*)

Thus, claim adherents to this view, is the problem solved of the wave-function collapse; it never occurs, but the different futures which the superposed waves decree are all pursued in different universes.

4.13 For some commentators the biggest challenge of the Copenhagen viewpoint is its *indeterminacy*, the fact that the future of a quantum system is not uniquely determined by its past history. There are, of course, many situations in the macroscopic world where we are unable to predict the future, but quantum indeterminacy is of a different nature; it is claimed that the data necessary for prediction just does not exist; quantum systems are intrinsically uncertain. Einstein was the most famous non-believer of this view. He maintained throughout his life that quantum theory was "incomplete"; there must be "hidden variables" which we had not yet discovered, and when we do understand them we shall, in principle, be able to predict the future of quantum systems as surely as we can preduct macro systems when we have the necessary resources.

4.14 Among those physicists who do broadly accept the Copenhagen picture, there is disagreement over the relative importance they attach to the wave-function part of the picture, and the particle picture that emerges when a measurement is taken. Roger Penrose makes it clear in his writings that for him the wave is closer to "reality" than the particle. He also stresses that the changing pattern of the wave-form of a particle, as described by Schrodinger's equation, is completely deterministic, and he claims that it has nothing to do with probabilities. He writes:

> Regarding Ψ as describing the 'reality' of the world, we have none of this indeterminism that is supposed to be a feature inherent in quantum theory -- so long as Ψ is governed by the deterministic Schrodinger evolution. Let us call this evolution process **U**. However, whenever we 'make a measurement', magnifying quantum effects to the classical level, we change the rules. Now we do *not* use **U**, but instead adopt the completely different procedure, which I refer to as **R**, of forming the squared moduli of quantum amplitudes to obtain classical probabilities! It is the procedure **R**, and *only* **R**, that introduces uncertainties and probabilities into quantum theory. (*The Emperor's New Mind*, 1989).

4.15 Now to present the writer's own views on these problems let us consider firstly the question of indeterminacy. What exactly would it mean if we did eventually discover, because the quantum world contains hidden variables of which at present we are unaware, that the world really is deterministic? It would mean simply that, given sufficient data and sufficient calculating power, we could use our knowledge of the past to predict accurately the future. But bearing in mind that the past and the future have no objective significance, and exist only in the minds of individuals in relation to their own "now", of what importance is this matter of being able to calculate the events on one side of this arbitrary dividing plane from knowledge on those of the other side? To us, of course, it would be of immense importance. The biggest handicap we carry is the fact that our memory acts only backwards in time; it would be wonderful (or so it seems until we think about it deeply) if we could also remember the future! But to the universe at large it would be of little significance. There can be no question of the future being *intrinsically* indeterminate, in the sense that it is still open to alteration in a manner that the past is not, as we tried to show in Chapter 2. It is merely that we cannot see the future because our memory is not looking that way. If you give to an engineer a drawing of *one half* of an automobile engine, would he be able to complete the other half? He would probably make a good attempt at it; there is not much doubt where the main components would need to be placed, but on the other hand he might make some fundamental error; he might assume the distributor was driven from one end of the inlet camshaft, while in reality it should be the exhaust camshaft. Good fun, but not really important. And equally fascinating, but of no more importance, is the question of whether the future half of the history of the universe could be deduced from its past history. So it seems surprising that Einstein, who surely understood the nature of time better than anyone, should be so concerned about the apparent indeterminacy of quantum theory. He would probably have relented had he lived to see the EPR experiments conducted in the 1980's, (which we discuss in Chapter 9), and the fact that even if the existence of hidden variables could remove the indeterminacy from quantum interactions, the *non-locality* revealed by these experiments would still remain.

4.16 Some aspects of Penrose's view are puzzling. He gives priority of importance to the wave-function because it does not involve probabilities. "It is the procedure **R** [the process of making a measurement], and only **R**, that introduces uncertainties and probabilities into quantum theory". But does this not overlook the fact that the wave is a *probability amplitude*? It was introduced by the quantum theorists in the 1920's to describe the very probabilities which Penrose says do not exist. And his interest in the fact that **U**, the wave function, evolves deterministically between one measurement and the next, seems misplaced. **U** represents our best attempt to predict the future behaviour of a quantum system using the knowledge we have of its past; if **U** changes with time other than deterministically, then it can not represent our best effort. Until we have more information, any need to alter the prediction we have made, on the basis of information we already have, can arise only because we have made a mistake.

4.17 But the most telling objection to all the viewpoints we have described above is that they are not time-symmetric. The behaviour of any collection of particles *is* reversible, provided they are few enough for thermodynamical considerations not to apply. As the waveform is in essence a *description* of the particle's behaviour, it too should evolve in a way that can be viewed in reverse without breaking its own rules. The collapse of the waveform, whether we regard it as happening when a particle collision occurs, or when a measurement is made, or when a conscious mind *knows* a measurement has been made, is certainly irreversible, and indicates to us that the wave function, as understood in these theories, cannot be a part of physical reality, and can exist only in the mind of the observer. And the splitting of the universe into two is even more grotesquely irreversible.

4.18 The one fundamental principle which lies at the heart of all the perplexities of the quantum world is the fact that a quantum system, when faced with a choice between two or more different ways in which to develop, seems to keep *all its options open* until it is next observed. But what does this mean? The very idea is based on an assumption that, while the past is immutable, the future is undecided, an assumption which we hope the arguments of Chapter 2 have dispelled. When we view the history of such a system as a picture in the four dimensions of space-time, there is no dividing line between the past and the future, and both are drawn with the same firm strokes of the pencil. It is quite without meaning to claim that the future remains open while the past is closed. When this is accepted it becomes clear that a belief in this doctrine cannot reflect anything in the world outside our own minds. The uncertainty in the future of a system, and its apparent tendency to keep its options open, can be no more than a reflection of our own limitations. This does not provide an escape from the dilemma, but it does show that any way out proposed by the doctrine of an immutable past and an undecided future, is in fact just a cul-de-sac.

4.19 In recent years a number of modified versions of the Copenhagen interpretation have been developed which do reveal much deeper understanding of the nature of time, and which acknowledge the need to describe "time-symmetric" phenomena by means of a picture which is itself time-symmetric. These deserve more attention than they seem to have received. John Cramer's "Transactional Interpretation" is carefully thought out, and he describes it in his writings with admirable clarity. He claims that it resolves some of the paradoxes which remain with the Copenhagen picture, and in particular the unique and questionable role performed by the *observer*. The essence of Cramer's model is the description of any quantum event as a "handshake", executed through an exchange of a normal "retarded" wave and an "advanced" wave, which in effect acts "backwards in time". Thus if points A and B on the space-time diagram (Chapter 2, fig. 2-5) represent consecutive events on the world-line of a particle, the retarded wave acts from A to B, and corresponds closely to the wave described by the Schrodinger equation. Then the advanced wave completes the transaction by acting in reverse from B to A. Cramer writes:

> This advanced-retarded handshake is the basis for the transactional interpretation of quantum mechanics. It is a two-way contract between the future and the past for the purpose of transferring energy, momentum, etc., while observing all the conservation laws and quantization conditions imposed at the emitter/absorber terminating

'boundaries' of the transaction. The transaction is explicitly non-local because the future is, in a limited way, affecting the past (at the level of enforcing correlations). It also alters the way in which we must look at physical phenomena. When we stand in the dark and look at a star a hundred light years away, not only have the retarded light waves from the star been traveling for a hundred years to reach our eyes, but the advanced waves generated by absorption processes within our eyes have reached a hundred years into the past, completing the transaction that permitted the star to shine in our direction. (*An Overview of the Transactional Interpretation* , International Journal of Theoretical Physics, 27, 227 (1988))

4.20 At first sight it appears that Cramer is making the mistake described in Chapter 2 and illustrated in Fig. 2-5. What is the difference between a wave starting at A and ending at B, and one starting at B and ending at A? We must remember that A and B are not points in space, but events in space-time, and that the movement of a particle from one point to another is represented simply by a *line* such as AB, and not by a point *moving* along AB. At each end of their trajectories the two waves occupy the same position and time; so they move *together* from one point to the other. Claiming that the advanced wave moves *backwards in time* is redundant and meaningless. But as one studies Cramer's writing it is clear that he understands this perfectly. The only sense in which one wave goes from A to B and the other from B to A is that A is *responsible* for the retarded wave, and B responsible for the advanced wave. This question of *responsibility* or *dependence* is very difficult to define, and it does seem possible that Cramer's model is unnecessarily complicated, and that indeed the advanced and retarded waves are different descriptions of the same entity. Even if this proves to be the case, however, the most valuable part of Cramer's description concerns the transference of *information*, such as momentum and energy, between consecutive events on the world line of a particle. The idea that such information is transferred by the wave, and not by the particle itself, will become an essential feature of the Nodal interpretation which is developed in the remainder of this book. Cramer's model is the first one we have examined which preserves the time-symmetry of the events it purports to describe.

4.21 Huw Price is another writer who believes that the mysteries of the quantum world can be explained if one is prepared to accept that, in some circumstances, earlier events can be influenced causally by later ones. Price is a philosopher, but his understanding of the subtle interpretation of physical principles is admirable, and every scientist with an interest in the fundamental nature of things should read his book *"Time's Arrow and Archimedes' Point"* (OUP, 1996). His analysis of the apparent "direction" of time, and of how one can avoid falling into the traps set by our deep-rooted intuitions concerning the flow and the asymmetry of time, is surely unsurpassed. Price believes in a form of "advanced action", whereby the state of an incoming quantum system, as represented by its "hidden variables", is influenced by the *future* setting of the apparatus which is going to perform a measurement. He shows persuasively that such a hypothesis can explain away the measurement problem, the superposition puzzle and its effect in interference experiments, and the EPR enigma, while preserving the idea of "locality", that influences can be transmitted (in either time direction) no faster than light. The incredulity with which we react to suggestions that causal influences can act backwards in time is, of course, evidence of the power of our deep-rooted temporal prejudices, and should not preclude a careful appraisal of theories like Price's. The micro-world is essentially time-symmetric; we accept that past events can affect the future, so surely there is no reason for us to reject the opposite, and Price mounts a robust defence of the idea. But is there a need for Price's theory? Its chief justification is that it abolishes the need for influences to transmit at super-luminal speeds, but why should this be so important? Einstein shows that neither *particles* nor *information* can travel faster than light, but the influences that are involved here are not of this kind, for they merely influence future *probabilities* rather than events themselves. We have no experience of such influences, and no direct way of knowing how they transmit, nor any reason to doubt that they can operate faster than light, so it seems to the writer that the "advanced action" hypothesis is ingenious but unnecessary.

4.22 Another book which handles the "arrow of time" problem with great insight is L. S. Schulman's *Time's Arrows and Quantum Measurement* (CUP, 1997). Schulman treats his subject matter more mathematically than Price, but he too resorts to a type of "backwards causation" to resolve the problem of the "collapse" of the wave function. When we believe the wave contains a superposition of different incompatible states, as in interference experiments, then the *future* measurement or observation we are about to make *suppresses* all the approaching superposed states except one; there is thus no need for a collapse to realise the one state that comes about.

4.23 These imaginative interpretations are to be welcomed. Contrary to popular belief, all the great discoveries of physics and mathematics are not made by deductive logical reasoning, but rather by flashes of imagination, which can then be supported deductively. Einstein was not the world's greatest mathematician, but had remarkable powers of insight and visualisation, and it was these that led him to Special and General Relativity, his explanation of the photoelectric effect and other great discoveries. Almost equally brilliant were the insights of Bohr, de Broglie, Heisenberg, Schrodinger and Dirac during the first half of the twentieth century, from which the highly successful formalism of quantum mechanics has sprung. During the second half of the century there have on the other hand been a number of brilliant mathematicians, but they have done little to solve today's problems in cosmology or particle physics, which are concerned more with interpretation than deduction, and demand intuition rather than logic. Indeed it has seemed that some of today's mathematicians, far removed from the observatory or particle laboratory, have believed themselves to be solving the problems of the universe when all they were doing was shuffling symbols on a sheet of paper. As we enter a new century, can we hope that a new era of imagination and insight is dawning?

Chapter 5: The Nodal Viewpoint

5.01 As we have shown, there are several perplexities which most current interpretations of quantum phenomena appear not to resolve. These all stem from the puzzling particle-wave duality of sub-atomic bodies, with the associated problems of Heisenberg uncertainty, and the apparent "collapse" of the wave when an observation is made. This chapter presents in outline the writer's new "nodal" perspective, and attempts to explain how it meets the challenges presented by these phenomena.

5.02 Chapter 1 put forward the view that any description of the quantum world should embrace two principles if it is to be accepted. Expressing these two principles more precisely, they are as follows:

1. Seen from the particle viewpoint, every fundamental sub-atomic process (with the apparent exception of the decay of the neutral kaon) is *time-symmetric*. This implies that, if the process is illustrated by means of a moving picture, and the sequence of frames is run in reverse, the behaviour will still appear compatible with all the laws of nature with which we are familiar. It follows that any model we adopt in order to *visualise* atomic phenomena, or any mathematical process we adopt to *describe* them, must also be time-symmetric. All the best-known current theories of quantum mechanics seem to fail this reversibility test.

2. A particle can have no influence on any other particle unless it encounters what we are calling a *node*. This may be a collision between two particles which subsequently rebound, or a collision in which the particles coalesce (sometimes called *annihilation*), or a splitting of a particle into two (sometimes called *pair production*). We can never observe a particle at a particular time and place *unless* it has a node there. The essential feature of the Nodal Interpretation is the proposition that a particle *does not exist* except at its nodes. We suppose that, while these nodes *do* have real existence, and precise positions in space and time, the particles we picture moving from one node to another do not exist. The trajectories which we imagine these particles to trace out between one node and the next have no counterpart in the real world.

5.03 The usual meaning attributed to the wave function of a particle is incompatible with this nodal viewpoint, for the wave is supposed to represent a probability distribution for the position of a particle while it is "in flight", and we are denying the existence of the particle in such circumstances. But with a slight change of emphasis a particle's wave function can continue to play a useful part in our description. The squared modulus of Ψ, integrated over a region of space at any given time, is said to give the probability that the particle lies within that region at that moment of time. Instead of this, from the nodal viewpoint we must say that, *if the particle's next node* happens to occur at that given moment, then the squared modulus gives the probability that the node lies within that region. So $|\Psi(\mathbf{r},t)|^2$ gives us, not the probability density that a moving particle occupies position $\mathbf{r}$ at time t, but rather the probability that, given that its next node occurs at the time t, its position vector will be $\mathbf{r}$. Because of the time-asymmetry of our memories and of our record-keeping equipment, we often know something of the past history of a system, but not its future behaviour, and we try to predict this behaviour. Usually we can calculate only probabilities, whose values, it must be remembered, are essentially subjective, depending on the degree of knowledge we have of the past, and it is here that the wave function provides such a useful tool, even though it reflects our own biased view of reality as well as reality itself. The Nodal approach insists that these probabilities do not refer to the particle's position as it moves from one node to the next, but only to the position of that next node.

5.04 The particles which formerly we regarded as solid objects moving from node to node, we now claim to be fictitious, while the wave-form, in both its subjective and objective parts, consists of nothing more concrete than *information*. The *nodes* are the only substantial parts of this model, and indeed we assert that they constitute the only "material" things in the universe. So we must ask what are the properties of these nodes, and how we should visualise them. Each one has a definite position in space-time, corresponding to exact values of x, y, z and t, and under rather special experimental conditions we can sometimes determine these values with whatever degree of accuracy our methods of measurement will allow. It is only when the experimental set-up could give us some additional information about the momentum or energy of the (imagined) particles which come together at the node, that the degree of precision is reduced by the Heisenberg principle.

5.05 Each node must act as a "clearing house" for information relating to the particles, for there are restrictions on the possible relative positions which nodes can occupy, restrictions which can be enforced only by the interchange of information. In particular the positions of nodes in time and space must ensure that the (imagined) particles satisfy the conservation laws, such as the conservation of energy and momentum, and they must also pass on to subsequent nodes information about particle spin, electric charge, and the other properties which in the past we have attributed to the moving particles themselves.

5.06 Since, in this picture, we are denying the existence of a particle between one node and the next, we can no longer define its velocity as *dr/dt*, nor may we picture its speed as the swiftness of its motion, comparable with that of macroscopic objects like cars or planets. And yet we must still keep hold of the notion of *momentum*, for not only is momentum conserved in microscopic collisions, but the magnitude and direction of a particle's momentum is closely related to the wave picture which we have developed to describe a particle "in flight", as given by de Broglie's formula. Whatever place the wave function eventually holds as we develop nodal theory, it is clear that the concept of momentum, and hence of velocity, must continue to figure in our theory. We discuss this in a later chapter, but for the present we can show that in simple cases it does not present a stumbling block. Suppose a particle has a consecutive pair of nodes in free space, in a situation where previously we would have considered it to move between them in a straight line at constant speed. Because the *x, y, z* and *t* co-ordinates of the two nodes have perfectly definite values, there is no difficulty in defining the velocity between the nodes as $(\mathbf{r}_2-\mathbf{r}_1)/(t_2-t_1)$, and this is the value which the nodes will "take into consideration" in ensuring that momentum is conserved there. It follows that, for a consecutive pair of nodes which lie in *the past*, there is no reason in principle why we should not know accurately both their positions in space and time, and the momentum with which the imagined particle leaves the first and "arrives at" the second. The Heisenberg principle does not apply in such cases; it was pointed out in Chapter 1 that there has often been confusion concerning the situations in which the uncertainty principle does or does not apply. Our new approach changes the way in which we think about momentum, but does not change in any way the knowledge we can have of the momentum of a particle, or any of the results we expect from experiments.

5.07 The best way to represent quantum processes in diagrams presents us with a problem. It is easy to show the time and position of nodes on a space-time diagram, omitting, of course, one or two of the spatial dimensions because we cannot visualise or portray more than three. But how can we illustrate the relationships between nodes? We have seen that the trajectory of a particle from one node to another is not real, and this would seem to rule out drawing the usual Feynman diagram. But in general each particle preserves its identity through a series of collisions, and we can show this only by joining together in our picture the successive nodes visited by a particle, and the lines we draw are necessary also to illustrate the way in which energy and momentum are transmitted through a system, and conserved at each node. So it seems best to join up our nodes with straight lines, so far as the experimental setup allows; but it must be remembered that a straight line on a space-time diagram represents a particle moving with uniform velocity, and this is not the picture we should have in our minds.

5.08 There are many situations in which the lines connecting nodes cannot be straight, for instance when the "path" of a particle is diffracted as it "passes through" a small aperture. How is the disposition of nodes determined by the probabilities associated with different angles of diffraction? How, for instance, does a photon "know" that it may be diffracted through a large angle if the size of the aperture is small compared with the wavelength of the photon, but that this is unlikely if the aperture is larger? It is clear that, in some way, the relative positions of consecutive nodes are partly determined by the layout of all the matter in their vicinity, and that "knowledge" of this layout must be immediately available, with no restrictions such as that limiting transmission to sub-luminal speeds. Indeed, because the notion of simultaneity is denied by Special Relativity, this knowledge must encompass a whole region of space-time, past, present and future. We have here a generalisation of the "non-locality" revealed by the EPR experiments, wherein pairs of "entangled" particles show correlated behaviour which cannot be explained by sub-luminal transfer of information, as we shall discuss in a later chapter. The existence of non-local influences has gradually come to be accepted by proponents of all the differing quantum interpretations, for the results of these experiments have consistently refused to be explicable in any other way. We see now that such influences really apply much more widely, playing a part in determining the dispositions of nodes in all normal circumstances, and not only in the rather special situations met in EPR demonstrations.

5.09 We must stress that this purveyance of information throughout the whole of space-time does not imply the possibility of faster-than-light travel, nor does it provide a means of foretelling the future, or travelling into the past. We are material beings, composed of particles (or nodes) and restricted by all the laws of classical physics which particles obey. Claiming that these particles exist only when they collide does not change the laws which they obey. Although we must now regard the conservation of energy and momentum, and the restrictions of Special Relativity, as properties of the inter-relation between nodes rather than restrictions on moving particles themselves, these laws still apply. Anything we want to move from one place to another, and any information we wish to transmit, must be carried by the (imagined) particles of matter or radiation which we picture moving from node to node. It is the disposition of these nodes which ensures that particles of light or matter are still restricted in the speeds they can attain.

5.10 It is interesting to compare this concept of nodes being "aware" of the whole surrounding region of space and time with Feynman's "sum over histories" picture. The value of this method lies not in its use for doing actual calculations, for in most situations it would prove more involved and tedious than the standard method using Schrodinger's equations, but in the remarkably clear picture it gives of the significance of these equations. J. C. Polkinghorne's description cannot be improved upon; he explains here the sum-over-histories approach to the well-known experiment in which a stream of particles passes through two slits and produces an interference pattern on a screen, as if it consisted of waves, not particles:

> ... Feynman tells us that we should think of all the different ways in which an old-fashioned electron with classically picturable simultaneous position and momentum could travel from the source through slit 1 and onto the specified point on the screen. There is obviously a vast number of such possible trajectories. ... Each is called a "path" or a "history" (both terms are used). Feynman tells us to consider all such possibilities and to assign to each a complex probability amplitude. It involves a quantity which physicists call action. ... Feynman next gives us a rule for associating a complex amplitude with this number. [For the learned it is exp(iA/h)] You then add together all the contributions from all the different parts and -- hey presto! -- the result is the same probability amplitude you would have calculated by the more pedestrian procedure of solving the Schrodinger equation. (*The Quantum World*, Penguin, 1984)

5.11 As an illustration of this method, Feynman shows that in open space a photon seems to travel in *straight lines*, and writes as follows. (In his simplified description, he uses *arrows* to describe the vectors or complex numbers which Ψ represents).

> For each crooked path, ... there's a nearby path that's a little bit straighter and distinctly shorter -- that is, it takes much less time. But where the paths become nearly straight, ... a nearby, straighter path has nearly the same time. That's where the arrows add up rather than cancel out; that's where the light goes. (*QED*, p.54).

This clearly parallels closely the concept of nodes "feeling out" the whole of the space around them where adjacent nodes may lie. Such an "awareness" of the disposition of material bodies throughout the whole of the surrounding space-time must feature in any theory which claims to account for simple interference effects such as those observed in the "two slits" experiments.

5.12 What interpretation must we assign to Ψ in our theory? It was pointed out in Chapter 1 how clearly Heisenberg realised that the wave function is an unhappy amalgam of two sets of ideas, one objective and one subjective. How can we separate the two? The wave-function is clearly *objective* in that it encapsulates the identity of the particles whose trajectories we visualise as lines connecting one node with another, even though we acknowledge that these lines can mislead us by suggesting particles actually *travelling* through space as time progresses. And it is objective also in representing the information transferred between nodes to ensure that all the conservation laws are satisfied by their relative positions in space-time. But it is *subjective* in that it presupposes a *moving* time, and it is grossly time-asymmetric, for it is determined by the past nodes of a particle but not those lying in the future, reflecting the time-asymmetry of our own mental processes. This is most clearly seen if we consider a photon scattered from a material particle such as an electron. We have no idea in which direction the photon's next node will lie, and our ignorance is represented by the expanding spherical wave which we imagine to radiate outwards from the source, only to "collapse" when we know the whereabouts of the next node. If, by some miracle, we could know the position of the later node, but not the earlier one, then our probability wave would have to converge on the former rather than radiate from the latter. This alternative situation is not as bizarre as it may seem, and has been considered in detail by some physicists; they refer to such converging waves as *advanced*, to distinguish them from the more familiar *retarded* waves of every-day experience. We shall devote a later chapter to consideration of these two kinds of radiation. For the time being we must be satisfied with the dictum that those aspects of the wave function which are asymmetric in time are subjective, and assume that those which are symmetric are objective. We can visualise the objective features of the wave as providing the information linkage between nodes, but our picture is still imperfect if it suggests that the information "travels" from one node to the next, and that there is a preferred time direction.

5.13 We shall quote Heisenberg again, showing his understanding of the dual nature of the wave function, the subjective and the objective. Writing about the motion of an electron, he wrote:

> The probability function ... represents a fact in so far as it assigns at the initial time the probability unity (i.e., complete certainty) to the initial situation: the electron moving with the observed velocity at the observed position ... It represents our knowledge in so far as another observer could perhaps know the position of the electron more accurately. ...

When the probability function has been determined at the initial time from the observation, one can from the laws of quantum theory calculate the probability function at any later time and can thereby determine the probability for a measurement giving a specified value of the measured quantity. (*The Copenhagen Interpretation of Quantum Theory*, ch.3).

5.14 We see that the wave function represents in part facts about the quantum world it purports to describe, but at the same time incorporates information about an observer's knowledge of this world, knowledge which forms the basis of the probability predictions he can make, but which is purely subjective in that different observers may use different functions to describe the same phenomena.

5.15 In what follows we shall often need to distinguish between the standard wave function of quantum mechanics, with its combination of subjective and objective features, and the more restricted objective, time-symmetrical wave, which contains no features pertaining to our knowledge of a system. Throughout the rest of these pages we shall refer to the former as the *conventional wave function* (which we abbreviate to CWF), and the latter as the *nodal wave function* (which we abbreviate to NWF).

5.16 It is important to have in the imagination some picture of the world which takes into account everything we do know about quantum theory. Such a picture may play no part in the calculations which predict the outcome of experiments, and nor will it affect any deductive argument we conduct in attempting to develop our theories. But it can act as a valuable stimulus to the imagination, and so can have a powerful influence on the *inductive* arguments we present to ourselves when thinking about these theories. Our picture must be time symmetric, for we have agreed that the sub-atomic world, and any description or representation we give of this world, must continue to make sense if the direction of time is reversed. But we know that the idea of a moving picture or a moving time is itself invalid; if we view quantum processes in this way we are imposing on the system something that is subjective and illusory. The only way to be sure we are not thereby reaching false conclusions is to picture a static four-dimensional world, and to strive constantly not to allow any change or movement to impose itself on that picture. On this framework we can visualise a distribution of nodes representing the whole history of the universe, or more realistically a small part of this history. We are very tempted to picture our "now" plane moving upwards, to show our imagined "flowing" of time, but we must suppress this false picture. Equally importantly, there can be no question that events above this plane are in any way less real, less substantial, or less determined, than those below it. The distinction between the past and the future is purely within our own brains, and exists only because we can remember the past but not the future.

5.17 We may picture our nodes connected together by lines, *not* to indicate the trajectories of the particles, but simply to represent *which* particles are involved at *which* nodes, and to display the imaginary channels along which information is shared. Here again we must not see the information as *flowing* along these lines, either in one direction or the other. The nodes are just *there*, and the information shared beween them is just *there* also.

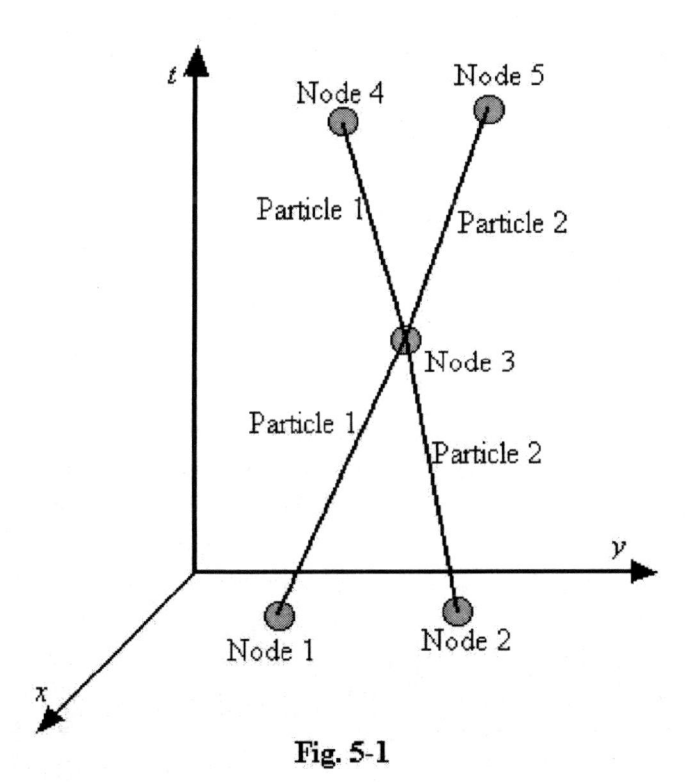

Fig. 5-1

5.18 The nodes are not distributed at random. An important part of the information which they share with one another ensures that their positions in space-time obey the laws of conservation of momentum and energy. It is instructive to examine in detail a simple situation such as a collision between two particles in free space. This is illustrated in the diagram, and it will be seen that *five* nodes are involved. The position of each one of these contributes something to the information exchanged at the node marking the actual collision, for we define the *momenta* of the particles arriving at node 3 in terms of the positions of nodes 1 and 2, and the momenta of the particles leaving node 3 in terms of nodes 4 and 5. To describe the positions of the five nodes would require twenty co-ordinate values (since our diagram represents a four-dimensional space). But these twenty values are not independent. We can write down three equations describing the conservation of momentum at node 3, and one for the conservation of energy. As a result the conservation laws reduce the number of degrees of freedom of the five nodes from 20 to 16.

5.19 We should examine further the effect of these conservation laws, for they throw light on one of the chief sources of quantum *determinacy*. Consider firstly a familiar problem in classical dynamics. If we are given a pair of smooth, perfectly elastic spheres colliding in three dimensional space, it is possible to calculate accurately the two velocities after the collision. We have six unknowns, the three components of the velocity of each sphere after the collision. And we have six equations; the conservation of momentum gives us three, conservation of energy gives one, and the fact that the relative velocity of the spheres in the plane perpendicular to the line of centres at the moment of collision is unchanged gives us two more equations. But in order to use this last fact we must know the orientation in space of this line of centres. Comparing this situation with the collision of two elementary particles, the conservation of momentum and energy again give us four equations, but there is no information analogous to the orientation of the line of centres at the moment of collision, which is meaningless for bodies of infinitesimal size. So we lack two equations, and cannot calculate the final velocities, however much we know about the state of affairs before the collision. We see that, at every collision of elementary

- 82 -

particles, from the six degrees of freedom which the final velocities possess, two of these are indeterminate.

5.20 If there is difficulty picturing a set of nodes, fixed in space-time, with some restrictions on their possible disposition, but with the information required to define these restrictions fixed also, and not being *transmitted* from node to node (which would involve some *change* in our four-dimensional picture), perhaps a little children's experiment in two-dimensions may help. Many of us have tried magnetising a number of needles, pushing these vertically into corks, and floating them in a basin of water. The needles repel each other, and they move into one of a number of possible positions of equilibrium. If we view this situation after the needles have taken up their final positions, their magnetic fields clearly influence their possible positions, but is not *moving* from needle to needle; in a comparable way, the nodes have fixed positions in space-time, with information imposing restrictions on their arrangement, but without any *flow* or *movement* of information being needed. Although this is an imperfect analogy, it may help us imagine a system of nodes in space-time, and their interaction.

5.21 But what corresponds in the quantum world with these magnetic forces in our model? What is the medium through which nodes influence each other? We attribute this influence to the Nodal Wave Function (NWF), but of its true nature we have no idea. Whether we shall ever understand the fundamental laws at a sufficiently deep level to answer this question is a matter for surmise and discussion. All we can do at the present time is to catalogue systematically the effects which these laws have on the nodes themselves. In the first place they decree that some arrangements of nodes in space-time are not allowed because they do not obey the rules of conservation of momentum and energy, and in a more complicated way, those other rules related to electric charge, angular momentum and phase information. Of those arrangements that are possible, some occur more frequently than others, giving us the probability element which our previous pictures of the quantum world have incorporated. We have seen that the nodal part of the wave function of a particle, the NWF, does really exist in some sense, and it is this which links together the nodes in any region of space-time, and "carries" the information which connects consecutive nodes along the (imagined) world line of a particle.

5.22 The reason behind the *Heisenberg uncertainty principle*, as it applies to position and momentum, can be clearly seen by considering a typical example. The only things we can observe directly are nodes, such as the collision of a photon with a photographic plate, and the only direct information we can gain from such an observation is the position in space and time of that collision. We cannot observe the wave form as it carries information from one node to another unless we introduce a new node, and we can discover facts concerning the energy or momentum of a particle as it arrives at our detector only by some kind of subterfuge, for example by introducing additional apparatus. And this additional information can make itself known to us only by *changing* the position or time of such a collision.

5.23 Suppose we want to know where a photon strikes an opaque screen, and find out also the direction from which it came (this latter representing *momentum* information). We could make a small hole in the screen, and place a photographic plate a small distance behind it.

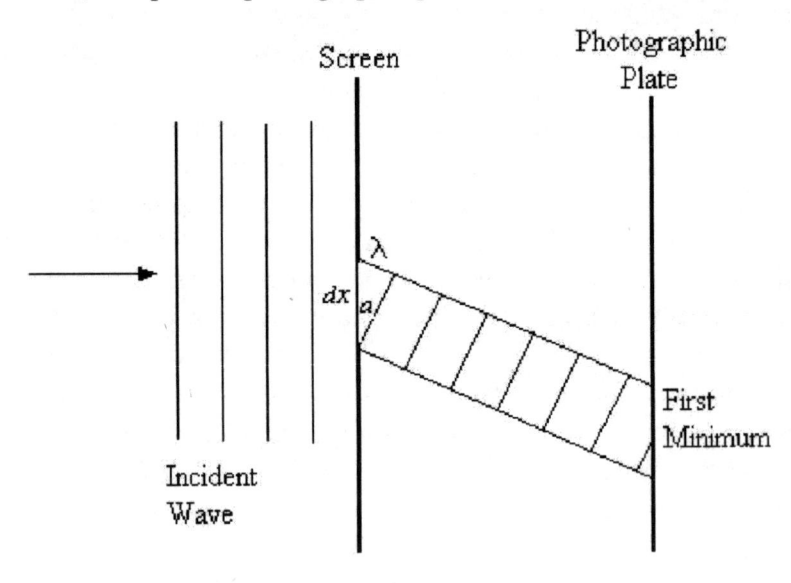

Fig. 5-2

When a photon "passes through" the hole we know its position as accurately as the hole in the screen allows, i.e. with an uncertainty of

dx, where dx is the diameter of the hole, and the mark it makes on the photographic plate should show us the direction from which it approached the hole. But in passing through the hole, the particle is likely to deviate from its straight course because of diffraction. In other words the momentum in a direction parallel to the screen, p, will suffer a change of unknown magnitude, dp, and this will introduce an uncertainty in our determination of momentum. A simple geometrical argument can give us the likely range of values this change will adopt, and shows that our uncertainty of the particle's position and momentum at the screen obeys the Heisenberg relation, $dx.dp > h$.

> [For those who want the mathematics:
> The probability that the particle reaches a particular point on the screen depends upon the angle a. The diagram shows the angle for which the first *minimum* occurs, when that part of the wave coming from the top of the aperture gets ahead of the part passing close to the bottom by one complete wavelength, and all possible phases are combined in the resulting wave. These will largely cancel each other, and we get a band of minimum brightness on the photographic plate, and the angle a can be taken as the characteristic angle of diffraction for this wavelength. We see that $\sin a = \lambda/dx$ and a simple "triangle of velocities" shows that $dp = p \sin a$. So $dp = p\lambda/dx$ or $(h/\lambda)(\lambda/dx)$. We derive immediately $dx.dp = h$.]

5.24 Diffraction effects such as this are very real; the pattern on the photographic plate confirms that they have occurred. So the waves we have envisaged at the hole in the screen must really exist; they belong to the NWF as well as to the CWF. We shall examine in detail the form taken by the NWF in simple cases in Chapter 11.

5.25 *Interference effects*, such as those produced by the famous "two slits" experiment can now be explained. There was never any difficulty understanding how these effects are generated by waves; they were discovered and explained long before the quantum was thought of. The problem lies in explaining them when a beam of light is treated as a stream of photons. If each photon were to pass through *both* slits it must divide into two; but this cannot be the case, for each would then have only half the energy, and twice the wavelength, of the incident photons, giving the wrong interference pattern. We now know the answer: the photons, regarded as particles, do not exist. They pass through *neither* slit; it is the wave which passes through both slits.

5.26 The troublesome *collapse of the wave* when an observation is made can now be viewed in a different light, when it is recalled that this aspect of the wave, because of its time-asymmetry, must belong to its subjective or conventional features. It exists only in the mind of the observer, and owes its form to our own ignorance of the future, an ignorance we can do nothing to relieve until we make another observation. It inevitably changes discontinuously when such a measurement results in additional information entering that mind. The "collapse" of the wave-function when we gain further knowledge of the system is no more surprising than the collapse of our anticipation when we learn that the winning number in a lottery is, or is not, the same as the number on our ticket. It must be admitted, however, that this is a rather facile dismissal of a problem which has troubled thinkers for a hundred years, and we shall attempt a more detailed analysis in Chapter 12.

5.27 Perhaps the most difficult concept which the Nodal Theory forces upon us is its non-locality. The information transferred between nodes must take into account the content of the whole of the surrounding region of space-time, including points which can not be reached at sub-luminal speed, and even points representing the future. But as explained above, the concept of non-locality has already been taken on board by all schools of thought in the quantum debate on the EPR phenomena. The writer believes that each of the other quantum interpretations stretches our credulity further than does this concept of non-locality, and herein lies the main reason for advocating the nodal view in preference to the others.

Chapter 6: Interference

6.01 The "two slits" experiment which Thomas Young conducted in 1801 appeared to show conclusively that light consisted of waves, in contradiction to the corpuscular theory which Newton had advanced a century earlier. It gave no indication of the nature of these waves, but it did allow their wavelength to be estimated. Young set up a monochromatic light source A to illuminate a narrow slit in a screen B, which in turn illuminates screen C containing two parallel closely spaced slits. The light passing through these slits then falls on a third screen D, and is found to display an interference pattern consisting of bright and dark bands.

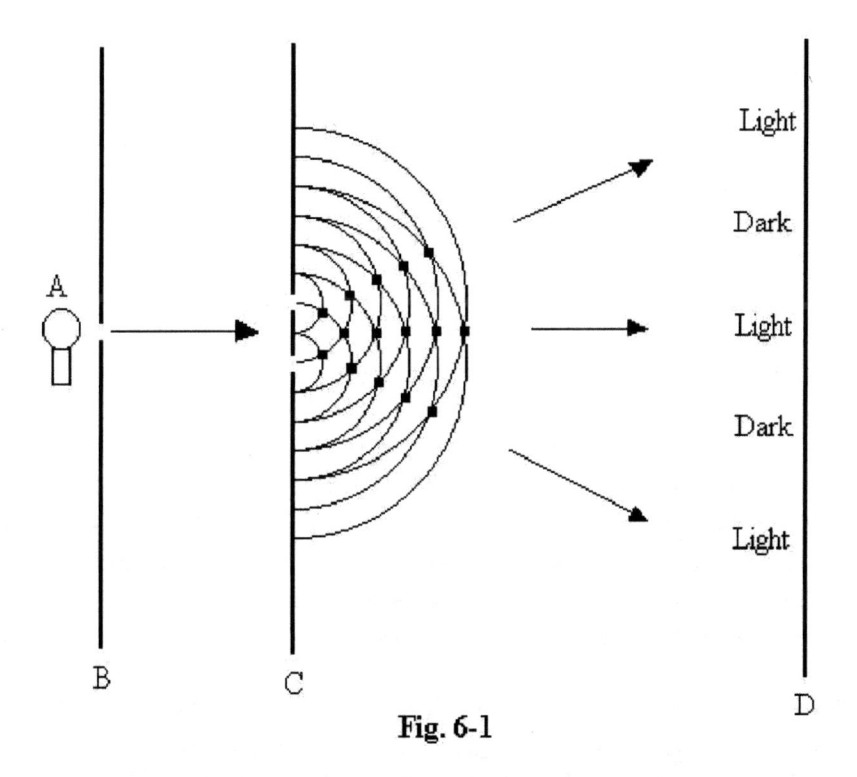

Fig. 6-1

The explanation depends upon the fact that at some points the waves from the two slits are in phase, and so reinforce each other to produce the bright bands, while at other points they are partially or

- 87 -

wholly out of phase. The circular arcs on the diagram show the wave crests radiating from the slits, and at points whose distances from the two slits differ by an integral number of wavelengths, such as those marked on the diagram, the two waves combine to produce a bright band, while at intermediate points the two waves annul each other to produce the dark bands. Simple geometric considerations permit calculation of the wavelength of the light, and show (very nearly) that the variation of brightness is itself sinusoidal, the points of maximum brightness having twice the average light intensity, and those of minimum brightness receiving no light at all.

6.02 By modern standards this is an easy experiment to repeat, and it seems to be just as easy to explain. But quantum mechanics has destroyed this apparently simple explanation, highlighting several of the most subtle problems which quantum theory has to face. That is why every elementary book on quantum mechanics devotes space to describing Young's experiment, and why we dedicate a chapter to it here.

6.03 After Louis de Broglie had suggested that particles of matter such as electrons might also display the same particle-wave duality as do photons of light, diffraction effects were soon discovered when electrons were scattered from crystals of nickel, or passed through thin metal foils. In theory it is possible to set up "two slit" experiments to show interference of electrons just as Young's experiment shows interference of light. Because the de Broglie wavelength of electrons at the speeds we can handle is very much less than the wavelength of visible light it is barely possible to do these experiments in practice, but from the behaviour of electrons as revealed in other experiments there is no doubt that the results would be the same in principle as those obtained with photons, with a sinusoidal electron distribution at the screen, the density at different points on the interference pattern varying between zero and twice the average density.

6.04 Schrodinger's equation threw some light on those phenomena which seemed to demonstrate the wave nature of the particles involved, whether of light or matter, and showed how interference effects could arise when particles have two or more possible routes to a particular point on the screen. The Ψ values describing

Schrodinger's waves were expressed as *complex* numbers, which can differ not only in magnitude but also in *argument* or *phase*. When added together such numbers can cancel each other because of their differing phases even when their separate magnitudes are not zero, so explaining the occurrence of the dark fringes in interference experiments. In this picture Ψ is a wave of *probability*, and the particles are still regarded as the principle actors on the stage, particles which can be detected individually in some experiments. So several questions were still unanswered. The interference pattern is produced only if both slits are open. When one slit is closed, some points on the screen receive *more* electrons; how do the electrons passing through the open slit "know" whether the other one is open? And when both slits are open, through which one does an individual electron pass? Or does it go through *both*, or *neither*? The Nodal theory provides an answer here, for it replies, "through neither", but a slight elaboration of the experiment seems to throw doubt on this. Suppose we shine a beam of light across the slits, so that we can observe any electrons which emerge from them. We find that electrons can indeed be seen coming through the slits, and each is found to pass through *one*, and only one, slit. If this were the whole story it would, of course, pose an even more impossible dilemma than the one the experiment is designed to settle, for the probability that an electron reaches a particular region of the screen D when both slits are open must equal the sum of the probabilities that it comes through the one or the other. No interference fringes could be observed, and in particular no region of D could experience *less* electron strikes with both slits open than with just one. But whenever we set up an experiment which would allow us to determine through which slit each electron passes, then the interference pattern *vanishes*. It is not difficult to understand what is happening. An electron can be seen only if it collides with at least one photon, with a random exchange of momentum and hence a change of the electron's wavelength. The wave which passes through the other slit suffers no such change, and so the two beams lose their coherence and interference is suppressed.

6.05 But this is where a multitude of "gee whiz" commentators took over, and did much damage to the understanding of quantum theory during the second half of the twentieth century. Even Richard Feynman, one of the most precise and accurate reporters despite his lively and popular style, was tempted to exaggerate. In the first sentence of the following quotation he almost succumbs to the temptation, but immediately corrects himself:

> If one looks at the holes or, more accurately, if one has a piece of apparatus which is capable of determining whether the electrons go through hole 1 or hole 2, then one can say that it goes either through hole 1 or hole 2. But, when one does *not* try to tell which way the electron goes, when there is nothing in the experiment to disturb the electrons, then one may *not* say that an electron goes either through hole 1 or hole 2. If one does say that, and starts to make any deductions from the statement, he will make errors in the analysis. This is the logical tightrope on which we must walk if we wish to describe nature successfully. (*The Feynman Lectures III*, p.1-9).

6.06 Other writers, less self-critical than Feynman, were variously attracted by the idea that a quantum system is *changed* when it is observed or measured, or when "we look at it", or "when the results of an observation enter someone's consciousness". Here is one example:

> The electrons not only know whether or not both holes are open, they know whether or not we are watching them, and they adjust their behavior accordingly. There is no clearer example of the interaction of the observer with the experiment. When we try to look at the spread-out electron wave, it collapses into a definite particle, but when we are not looking it keeps its options open. (*In Search of Schrodinger's Cat*, John Gribbin, p.171).

This is very misleading, and statements like this must bear some of the responsibility for such extreme interpretations as that of John Wheeler, who suggests that nothing in the universe really existed until there were conscious beings capable of looking at it. He writes,

> Is the very mechanism for the universe to come into being meaningless or unworkable or both unless the universe is guaranteed to produce life, consciousness and observership somewhere and for some little time in its history-to-be? (Quoted by Paul Davies, *Other Worlds*, p.126)

6.07 The nodal picture is much less dramatic. When interference is observed, only the wave function "passes through" the slits; the electrons do not exist there. But when we turn on the light to see the electrons as they emerge from the slits, and a photon strikes one of the electrons, then we have created a new node. The electron engages in an exchange of information there, and we know which slit it has passed through, while the momentum relayed from this node to the screen will not be the same as the original wave would have contained. The two parts of the wave are no longer fine tuned to interfere with each other, and the point where the particle strikes the screen is not constrained by the interference conditions.

6.08 To earlier experimenters and theorists it seemed that Nature was cunningly preventing us from discovering through which slit a particle passes whenever that particle is contributing to an interference pattern. But is there no way in which we can outwit her? When that particle is an electron, and we try to detect its presence by shining photons of light on the slits, the resulting collisions introduce random momentum changes which destroy the coherence of the two waves at the detector screen. But suppose we reduce the intensity of the light (and increase the sensitivity of our optical equipment). If this reduction proceeds gradually, we shall indeed see interference fringes beginning to appear as the light intensity is reduced. However, this is easily explained by the fact that there are then fewer photons. Those photons involved in collisions each have the same momentum as previously, and the electrons with which they collide are affected to the same extent, so making no contribution to the interference pattern; it is those which escape collision which continue to build up this pattern, and as the light intensity is reduced the proportion of electrons which escape increases.

6.09 Another possible approach might be to increase the *wavelength* of the light which we shine on the slits instead of decreasing the intensity. The momentum of a photon is given by h/λ, so that the magnitude of the disturbance which the electron suffers on collision is inversely proportional to the wavelength of the incident light. This is indeed confirmed in practice, and as the wavelength increases the interference fringes do begin to appear on the screen. But at the same time it becomes increasingly difficult to form a clear image from the light scattered by the electrons, which we are examining by means of optical equipment. As is well-known, no optical arrangement allows the examination of detail smaller than the wavelength of the light with which we illuminate it, so as the interference returns we lose the ability to distinguish the position of one slit from that of the other. Nature wins again. It seems that our inability to track electrons which are contributing to an interference effect is not due simply to our poor experimental technique; some fundamental law of nature appears to be involved. This law is, of course, the Heisenberg Uncertainty Principle.

6.10 Looking at the mathematics, we see that the formation of interference fringes is always associated with the requirement that we add together two or more complex wave functions (or *probability amplitudes*), with their inherent phase components. And when interference is suppressed we find it necessary instead to add separate *probabilities*, which are real numbers containing no phase information. Feynman gives us a simple rule for determining which of these procedures we should follow. If two possibilities lead to final states which are fundamentally *indistinguishable*, then we must add Ψ values, and find the resulting probability by calculating $|\Psi|^2$ at the end of the calculation. But if we can distinguish between the two possibilities, we must calculate separately the probabilitity, $|\Psi|^2$, for each possibility, and add these values in the same way that we always add probabilities for mutually exclusive events. In Feynman's words,

> If you could, *in principle*, distinguish the alternative *final* states (even though you do not bother to do so), the total, final probability is obtained by calculating the *probability* for each state (not the amplitude) and then adding them together. If you *cannot* distinguish the final states *even in principle*, then the probability amplitudes must be summed before taking the absolute square to find the actual probability.

Applying this to the present case, if the photon has a wavelength much smaller than the distance between the two slits, so that by examining the scattered light with a microscope we could decide which slit the electron has passed through, then we treat the two possible routes of the electron as real events, and add the probabilities. The distribution of electrons at the screen D then shows no interference effects. But if the photon's wavelength is considerably larger than the slit spacing, the scattered photons do not enable us to determine sufficiently accurately where they originated, and so we add Ψ values for the two slits. The *phase difference* between the two waves as they arrive at the screen will then be significant, and interference will occur.

6.11 If we imagine the experiment being repeated many times, each time with a slightly shorter photon wavelength, we know that the interference is *gradually* lost, and we see that Feynman's rule breaks down as we pass through the critical values where *partial* interference occurs, for in every case Feynman tells us we must *either* add probabilities *or* Ψ values. At what point are we to cease applying one formula and adopt the other? It may be thought that the Nodal interpretation suffers from the same defect. We explain the loss of coherence resulting from collisions with photons as being due to the extra node which the photons introduce into the situation; but as we gradually decrease the photons' wavelength, and so increase their energy, at what stage do we decide that the previously non-existent node has come into being? In replying, we must remember that the *only* function of a node is to exchange information at a particular time and place; the question of a node's "existence" is not important. Consider an analogy: if you ask whether a certain Mr. Smith "exists" then the answer must be just "yes" or "no", but if you ask whether Mr. Smith has influenced your life, a yes-no answer is of little use; the answer may well be one of degree. It is possible that Mr. Smith has substantially changed the course of your life, or that he has had only small effects on it. Asking whether Mr. Smith's influence "exists" is pointless when the real question is the magnitude (and direction) of that influence. So it is with the existence of Nodes. Whether we say that a particular node "exists" is just a matter of convenience; any significant statement concerning a node must describe the type and magnitude of its influence on neighbouring nodes. It is this influence which increases gradually as the photon wavelength is reduced, and the interference is found gradually to disappear.

6.12 The interpretation of interference effects can often be accomplished most easily by considering the transfer of *information*.

We have shown that an atom is prevented from contributing to interference if a colliding photon takes away sufficient momentum to destroy the coherence. Indeed, careful consideration of the possible transfer of momentum which can occur will often provide an accurate indication of the degree to which interference effects will be observed. Now momentum is a part of the information carried by the wave function of a particle. Its wavelength is directly related to that momentum, so it is not surprising that tracing the way this information is distributed after a collision or a series of collisions will tell us much about potential interference effects.

6.13 The Nodal interpetation takes this argument further. Between one node and the next the wave contains the *only* record of the particle's momentum. And we claim that the nodes consist only of the tranferrence of information which occurs there. We are deliberately ambivalent about the degree of "reality" we ascribe to the NWF, but there can be no doubt about the reality of the information. Perhaps the universe consists only of information.

6.14 Nodal theory provides its own replies to each of the other problems which interference phenomena present. The answer it provides to the duality question, whether electrons and photons and other particles consist of waves, or particles, or both, or neither, differs substantially from the explanations offered by other theories. The particles themselves do not exist, but the points where they collide do. As every theory must agree that the particles cannot influence our senses or measuring apparatus *except* at points of collision, the observations we make of any system must inevitably be exactly the same as if the particles did follow continuous trajectories, so it is not surprising that all previous theories have supposed their existence to be real. The information linking one node to another also exists, and the NWF is our picture of the channel through which it is transferred. Our own prejudiced position, knowing about the past but not the future, complicates the *probability* role of the waveform, but if we separate out the elements which are subjective, the information which remains must continue to involve an element of probability, which we investigate in Chapter 11. We know that, given the positions of all the *past* nodes of a system, there is not sufficient information to determine uniquely the positions of future nodes, but some positions are more likely than others, and this reflects the probability element in the linkage. Notice that exactly the same would be true if we regarded time to flow in the opposite direction. Given the positions of all *future* nodes of a system, there would not be sufficient information to determine uniquely the positions of past nodes, and a different set of probabilities would be involved.

6.15 When we are asked which slit the photons pass through in Young's experiment we answer "neither". But the wave has access to the whole of space-time, and so it "knows" whether or not both slits are open. It is useful to picture the wave as passing through the two slits, although we must remember that this may be *no more* than a picture. The wave consists only of information; provided the necessary information is available at each node it does not matter how it gets there. The railway timetable giving information about trains from London to Edinburgh does not need to make the journey itself.

6.16 So the loss of interference in particle experiments, when attempts are made to determine the paths of individual particles by illuminating them with a beam of photons, is explained in the nodal theory in terms of information loss. A complete understanding is provided if we consider the transfer of momentum, remembering that momentum is not to be thought of as a property of moving, massive particles, but as a component of the information linking adjacent nodes to each other. Bearing this in mind, together with the fact that some recent experiments have involved a subtle *selection* of particles to produce interference, or to restore interference effects which had been supressed, should remove the element of paradox which some researchers have suspected.

Chapter 7: Momentum

7.01 At first sight several features of the Nodal interpretation of Quantum Mechanics seem to threaten the principle of momentum conservation, which is generally assumed to be a universal law of nature. We hope to show in this chapter how these difficulties can be overcome, and we will analyse the momentum considerations in a few typical situations.

7.02 Momentum enters into the Nodal picture in two guises, neither of which bears any likeness to the momentum of the classical physicist. The characteristics of momentum in Newtonian physics are motion and mass, the sort of things we are aware of if we are struck by a moving object, or are trying to move one that is stationary. In Nodal quantum physics the idea of moving particles plays no part, while mass is no more than a component of the information which forms the linkage between nodes. Furthermore, as we have shown, the concept of velocity is now abstract, being no more than the result of a division sum; in simple cases where we can think of particles "moving" from node to node in straight lines, we obtain velocity by dividing the distance between a pair of nodes by the time separating them. The momentum of the particle is found by multiplying this result by the particle's mass. The alternative representation is no less abstract. The wave-function which we imagine spanning the interval between these two nodes has an oscillatory character, and de Broglie's rule shows that the wave-length of the oscillations is a direct measure of the momentum of the particle it represents; in fact momentum = h/λ, where λ is wavelength and h is Planck's constant. The Nodal theory does not specify what degree of physical reality we should ascribe to the wave-function, but its usefulness in calculation is unquestioned.

7.03 Despite the different nature of momentum in the Nodal theory, the way in which the concept arises in the discussion of quantum processes is often very "classical". An early example was provided by the Compton effect, discovered by A. H. Compton in 1923. When X-ray photons are scattered by the free electrons in a block of graphite, the scattered photons are found to have their wavelength (and hence their momentum) changed to an extent that depends upon the angle of deflection. The results confirm that energy and momentum are both conserved as they would be in the collisions of elastic macroscopic bodies. Niels Bohr wrote about the effect as follows:

> This phenomenon afforded, as is well known, a most direct proof of the adequacy of Einstein's views regarding the transfer of energy and momentum in the radiation process; at the same time, it was equally clear that no simple picture of a corpuscular collision could offer an exhaustive description of the phenomenon. Under the impact of such difficulties, doubts were for a time entertained even regarding the conservation of energy and momentum in the individual radiation process; a view, however, which very soon had to be abandoned in the face of more refined experiments bringing out the correlation between the deflection of the photon and the corresponding electron recoil. (*Discussions with Einstein,* 1949).

7.04 The first of our problems arises as soon as we consider in more detail the simple "motion" of a particle from one node to the next in free space. If we know the particle has been emitted at a point A, but have no additional information, from our point of view the wave-function will radiate spherically from A. But the other particle involved in the collision at A must suffer a recoil equal and opposite to the momentum change in the emitted particle. How can this happen if the direction of this momentum is not yet determined? And if the emitted particle travels a great distance before its next collision, its momentum could be indeterminate for an extended period of time.

7.05 Einstein and Bohr considered this problem in their historic discussions. Thus,

> ... Einstein stressed the dilemma still further by pointing out that the argumentation [of quantum theory] implied that any radiation process was 'uni-directed' in the sense that not only is a momentum corresponding to a photon with the direction of propagation transferred to an atom in the absorption process, but that also the emitting atom will receive an equivalent impulse in the opposite direction, although there can on the wave picture be no question of a preference for a single direction in an emission process. (*ibid.*)

7.06 The Nodal picture appears to exacerbate this problem. Not only is the direction of the particle's "trajectory" unknown, but we are claiming that the particle *does not exist* between its emission and absorption, and so its momentum must be missing from the system during this period. If it is (mistakenly) assumed to *travel* from A to B, then the centre of gravity of the whole system undergoes no change of velocity. But if we deny the existence of the particle between A and B, then both at the instant of emission and the instant of absorption, the momentum of this complete system suffers an instantaneous change, and between these two times momentum appears not to be conserved. Like so many problems, however, this one vanishes as soon as it is looked at from the correct viewpoint. We should be viewing the interaction of these particles as a static four-dimensional pattern of nodes in space-time; any picture we have of particles "moving" on this pattern, carrying momentum with them, must not be allowed to affect our understanding of the real world. The conservation of momentum is not concerned with the interactions of moving massive bodies; rather is it a rule restricting the possible positions of nodes in space-time. Some apparently possible dispositions of nodes do not satisfy these restrictions, while other do. All the arrangements which we find in practice are such that momentum is conserved. As the only information we can gain from a system is obtained by observing the space-time positions of these nodes, we never observe breaches of the conservation laws, for these occur only *between* nodes; the deficit existing during the "travel" of a particle is always restored on its "arrival".

7.07 The Heisenberg uncertainty does not prevent us from calculating accurately, in principle, the momentum of a particle between any pair of nodes lying in the past. There is no element of intrinsic uncertainty in the location of the nodes themselves in space-time, and, when referring to *past* collisions, there is no limit to the accuracy with which we can measure their positions and times of occurrence, and hence no limit to the precision with which we can describe the momenta of particles between collisions. We give two illustrations of methods by which the positions of a pair of past collisions can be determined experimentally to whatever degree of accuracy our techniques allow:

1. Suppose we wish to know accurately the points of emission and

absorption of an electron in a vacuum tube. We can release a brief burst of electrons from position A at time t_A. Then if one of these is detected by a flourescent screen at B, there is no fundamental restriction on our knowledge of the positions A and B, or of the times t_A and t_B. So the velocity between A and B can be found, and the momentum determined.

2. A different procedure must be adopted for photons. Unlike an electron, a photon does not possess a rest-mass, but on the other hand, we do always know its speed of travel (using the old language of "moving" particles). Limiting the duration of the emission of a burst of photons would lead to uncertainty in their energy, and hence their momenta, because of the Heisenberg effect. But in this case we can use a source of known wavelength, and allow a *long* burst to be emitted to minimise the uncertainty. When one photon is detected at B at time t_B we can use our knowledge of the velocity of light to calculate the exact time t_A of emission, and so again we can find accurately the position and time of this photon's emission and absorption.

7.08 Perhaps more troubling to the quantum physicist whose training was rooted in the classical tradition, is the realisation that the wave-function radiates in all directions from the emitting atom, and yet the photon which it represents seems to "know" in which direction to go in order to reach its next node, for this direction must be opposite to the emitter's recoil. We must remember, however, that the "radiation" of a wave-form is merely a picture necessitated by our prejudiced viewpoint; we may know the location of event A, but until B occurs we cannot know its location or time. The "real" wave between A and B, the NWF, cannot incorporate this feature of *radiating* from A, for it must be time-symmetric. If our mental moving picture of what is happening looks less reasonable when run in *reverse*, we know it must be wrong. In fact the NWF "knows" the location of B, and so does not need to radiate outwards searching for it. We shall consider the apparent asymmetry of radiation in greater detail in Chapter 8, and in Chapter 11 we will examine the general form taken by the NWF in some simple situations.

Fig. 7-1

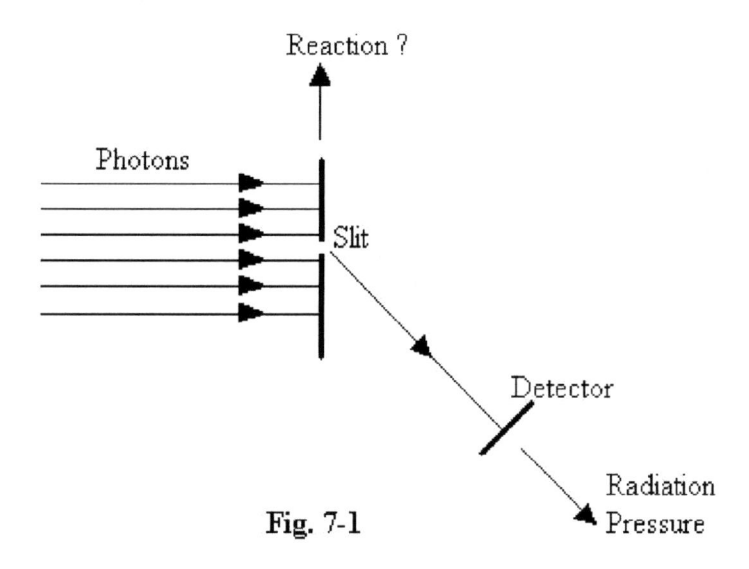

7.09 The next problem we consider occurs with an experimental arrangement in which photons of light fall upon an opaque diaphragm containing a narrow slit. The emerging light will be diffracted, and if a light-sensitive device is placed off-centre to observe photons which have apparently changed direction on passing through the slit, we seem to have a momentum paradox. When the experiment is conducted macroscopically, involving a large number of photons, the detector will clearly receive all its light from the direction of the slit, and will experience an impulse due to radiation pressure. This can not be balanced by a reaction on the *light source*, for all the photons which reach the detector have started their "journey" in the direction of the slit, and their reaction on the source must be opposite to this. The classical explanation demands a reaction between the photons and the *diaphragm*, for it is here that they change direction in passing through the slit. Bohr and Einstein considered this question in detail in their discussions in 1927. From the record which survives of these discussions it seems that both men assumed implicitly that indeed there is a transfer of momentum between the photon and the diaphragm; their disagreement was only about the possibility of *measuring* this transfer. Einstein at first was sure such a measurement would be possible, but had to accept Bohr's argument that it is not. Bohr shows that the uncertainty principle must be applied to the diaphragm as well as to the particle. He writes,

Here, it must be taken into consideration that the position of the diaphragm has so far been assumed to be accurately co-ordinated with the space-time reference frame. This assumption implies, in the description of the state ... an essential latitude as to its momentum. However, as soon as we want to know the momentum ... [of the diaphragm] with an accuracy sufficient to measure the momentum exchange with the particle under investigation, we shall, in accordance with the general indeterminacy relations, lose the possibility of its accurate location in space. (*ibid.*)

7.10 The Nodal picture of the situation is different. Firstly it must be pointed out that the diffraction effects are not produced by photons colliding with the *edges* of the slits; if they were, the resulting distribution of light, and its dependence on the width of the slit, would both be entirely different from what is observed. The diffraction is due to the diaphragm *shielding* the off-centre regions from any direct radiation, radiation which, if present, would cancel the diffracted beam, as described so well by Feynman's "sum over histories" picture. So if we consider a single photon, it is clear that it cannot suffer a collision at the slit; only the wave passes through the slit, and its change of direction is due to the absence of the waves on either side which would neutralise any wave moving in the "wrong" direction. Such a photon therefore cannot transmit any momentum to the diaphragm since it does not have a node there. Indeed the Nodal theory denies that it passes through the slit; it does not exist between the source and the detector.

7.11 But does this not lead to a contradiction in the case of the macroscopic experiment which we described at first, where a beam consisting of very many photons falls on the slit, and the light-sensitive detector placed off-centre experiences a light-pressure with a sideways component from those photons which are diffracted in its direction? No, there is no contradiction, because the photons which actually strike our detector are an unrepresentative selection of the set of photons which "pass through" the slit. In fact photons are diffracted to the right and to the left in (very nearly) equal numbers; the sideways component of the force on the off-centre detector will be balanced by the force exerted by those photons which are deflected in the other direction, and impinge on some other part of our apparatus. The nodal theory predicts here another microscopic breakdown of the conservation law, but of a magnitude which is essentially unmeasurable.

7.12 The above argument, that a photon deflected by diffraction will not exert a reaction on the apparatus which causes the diffraction, can not be applied, however, when a ray of light is reflected from a mirror. Here the change of momentum of all the photons is in the same direction, and can therefore be detected macroscopically; a stream of photons does exert a measurable force on a mirror which changes its direction. This can only occur if the photons experience real collisions with the particles constituting the reflecting surface. The question to be answered now is how these photons can retain their coherence despite these collisions. We saw in the previous chapter that the interference is lost in "two-slit" experiments if the photons suffer collisions as they pass through the slits. Interference effects can also be observed when a beam is split by a half-silvered mirror, and the two halves are brought together again by two fully silvered mirrors; why does the collision which a photon experiences at such a mirror not influence its coherence in the same way as is suffered by photons which are detected going one way or the other in two-slit experiments? The reason can be found in the fact that the molecules of a rigid solid such as a mirror are held tightly together by the rigidity of the solid. To act as a (theoretically) perfect mirror a surface must be infinitely massive and rigid, for it reverses the normal component of the momentum of the incident photons without affecting their energy, and so can not experience any recoil movement. It follows that the momentum of each photon can be changed in direction without any change of magnitude. A single photon gives up no *information* to such a fixed mirror, apart from that relating to the change in direction of its momentum, and so is still able to produce interference effects. Just as in the comparable two-slit experiments, a stream of photons which has been spilt in two can be brought together again by mirrors to cause interference effects. As in the two-slit case, it is impossible to determine which path an individual photon follows without at the same time destroying the interference.

7.13 But this requires further consideration. In the two-slit case we asked whether each individual particle passed through *both* slits, through *neither*, or through just *one*. The Nodal theory gave a clear answer: they pass through *neither*. In the case of mirrors, however, each photon must experience a collision to change its momentum, and so has a node at the point of reflection. Each photon must follow just *one* of the two possible routes through the equipment. The associated wave-form is "aware" of both routes, and it is the interference of the two waves which ensures that the photon arrives at a particular point of the interference pattern with the required probabilities. One mirror reflects just the wave-form itself, while the other introduces an additonal node, and a real collision of particles. No phase information is lost at either mirror, and as it is impossible to discover which route a particular photon has followed, the situations really are virtually symmetrical, with both routes contributing equally to the information delivered at the screen.

7.14 It is easy to see why interference will be lost if a successful attempt is made to determine which of the two paths the photon really follows, by looking for the recoil of one of the mirrors. The mirror must be able to detect a momentum change of the order of $p = h/\lambda$, and so it must be free to *move*. By the Heisenberg principle its position is indeterminate to within h/p, i.e. the photon's wavelength, and this uncertainty destroys the interference effects. Both the momentum of the photon and the length of its path are changed, so that the coherence of the two waves is lost.

7.15 We have examined just a few of the implications for nodal theory of the transfer of momentum in atomic experiments, and have had some surprises. In particular, the conservation of momentum, long held to be a universal law of nature, is found to be broken on the scale of single atomic particles, but only to an extent that is fundamentally unmeasurable and undetectable, and in situations which cannot result in *macroscopic* deviations from conservation. This explains why we have never encountered such discrepencies in experiments, and ensures we shall never do so.

Chapter 8: Advanced and Retarded Radiation

8.01 We have seen that some of nature's processes are reversible in the sense that, if time were to "flow" in the opposite direction, they would still appear plausible, while others become laughably silly if we think of them in reverse. All the simplest of nature's activities, such as the elementary interactions of atomic particles, and the movement of molecules in a gas, are essentially reversible, but more complex processes, those involving large numbers of particles, such as the flowing of rivers and the breaking of eggs, are usually not. We decided that this temporal asymmetry was not due to any of nature's fundamental laws, but rather to the very special state of the universe at the present time, a state of low entropy, with significant temperature differences and gravitational instability. This state of affairs must be ultimately due to the *boundary conditions* existing immediately after the big bang, conditions which, until we know their underlying reasons, seem highly unexpected. If the universe had started out in what seems to us a more reasonable state of randomness and disorder, then long ago it would have reached a state of equilibrium, with all the matter condensed into one gigantic mass or black hole, or with everything at the same temperature, so that nothing of any significance could ever happen.

8.02 One familiar process which appears to be irreversible is the *radiation* of energy. We see examples of radiation when waves spread out from a disturbance on the surface of a pond, when sound or light are generated by a loudspeaker or a lamp, or when waves are transmitted by a radio station. In each case, if we try to imagine the process taking place in reverse, the picture is so strange that we are convinced it could not occur. In his book *Time's Arrow and Archimedes' Point* (OUP, p.49), the philosopher Huw Price paints an attractive picture of an otter jumping into a still Scottish loch:

> ... we all know what happens, at least in one respect. Circular ripples spread outwards from the otter's point of entry, across the water's surface. It turns out that this familiar and appealing image illustrates another of the puzzling ways in which nature is asymmetric in time...

> The asymmetry turns on the fact that the ripples always spread *outwards*, rather than *inwards*. ... It makes no difference whether the otter is entering or leaving the water, of course!

8.03 Inward moving waves, if they ever occur, are often called *advanced* because they seem to exist *before* the disturbance which is responsible for them, to distinguish them from *retarded* waves, the more familiar type. Several scientists and philosophers have devoted much attention to considering these advanced waves, asking whether they could ever exist, and if not, why not. If such waves were ever to be observed, there would be two different ways of describing them. We could suppose that, like retarded waves, they are nevertheless still *generated* by the emitter, the lamp, the loudspeaker or the otter, but that they travel *backwards in time*; the emitter causes the radiation, as we expect, but the radiation exists before its cause. The alternative description pictures the emitter as the destination or absorber of the radiation; here we are faced with the problem that all the radiation must have been generated at a large number of separate points, for instance all the points around the edge of the loch inhabited by our otter, and the times of generation must have been finely co-ordinated to ensure that it all arrives at the emitter at exactly the right moment. Neither of these pictures appears reasonable, and the problem is equally acute when we consider radiation of light, radio waves or sound.

8.04 Why has so much thought been dedicated to this question? After all, is not radiation just one more from the wide range of phenomena which are uni-directional because of the Second Law of Thermodynamics? Is it not merely an example of the *dissipation* which results from the low entropy state of the universe today? We are not perplexed to the same extent by the irreversibity of falling stones and breaking eggs; these effects are not so mysterious that we find ourselves talking about the possibility of *effects* existing before their *causes*, or going backwards through time, as we must do in discussing advanced waves. Why do we single out this particular example of the asymmetry of nature?

8.05 Perhaps the explanation lies in the work of James Clerk Maxwell, to whom we owe the electromagnetic explanation of radiation, and who is, quite rightly, still held in admiration by many of today's physicists. May we quote from Huw Price's clear description:

> Maxwell's theory of electromagnetism, developed in the mid-nineteenth century, is easily seen to admit two kinds of mathematical solutions for the equations describing radiation of energy in the electromagnetic field. One sort of radiation, called the retarded solution, seems to correspond to what we actually observe in nature, which is outgoing concentric waves. The other case, the so-called *advanced* solution, describes the temporal inverse phenomenon -- incoming concentric waves -- which never seem to be found in nature. Thus the puzzle of temporal asymmetry here takes a particularly sharp form. Maxwell's theory clearly permits both kinds of solution, but nature appears to choose only one. (*ibid.*)

This point has been made by many writers, but does it not show a misunderstanding of the relationship between a mathematical equation and the phenomenon it describes? When we devise an equation to represent a real problem we are asserting that any resolution of the problem will correspond to a solution of the equation; we are *not* asserting that any solution to the equation corresponds to a resolution of the problem. Try this little problem in elementary algebra:

> You drive five miles to work each morning through heavy traffic. If you could travel 10 mph faster your journey would take five minutes less. How fast do you normally travel?

You should derive a quadratic equation, and find the answer to be either +20 or -30 mph. But we hope you will not set off tomorrow at 30 mph in the wrong direction, nor wonder why it is impossible to drive a car at 30 mph in reverse gear. We reject the *advanced* solution to the equation because it is not applicable to the problem. And we reject the advanced solutions to Maxwell's equations for the same reason.

[For those who want to see the mathematics:
Suppose usual speed = v mph.
So usual time = $5/v$ hr. = $300/v$ mins.
Time at faster speed = $300/(v + 10)$ mins.
So $300/v = 300/(v + 10) + 5$
Solving $(v + 30)(v - 20) = 0$
So $v = +20$ or $v = -30$]

8.06 And yet the problem of advanced radiation has indeed attracted a great deal of attention over the years, both theoretical and experimental. The best known attempt to show theoretically why we do not witness light sources radiating into the past was devised by Richard Feynman and John Wheeler in 1941. They considered the consequences of assuming that retarded and advanced waves *are* indeed generated by every source of electromagnetic radiation, so that a radio station, for example, sends half its power into the future and half into the past. They came to the surprising conclusion that all the advanced waves would disappear, and we would be left with what we had always believed, that the whole power of the transmitter is directed into the future. Paul Davies describes their reasoning well:

> When the retarded waves from a particular point on Earth, having spread out across the universe, encounter matter, they will be absorbed. The process of absorption involves the disturbance of electric charges by the electromagnetic waves, and as a result secondary radiation is produced by these faraway charges. The radiation too is one-half retarded and one-half advanced, in accordance with the assumption of the theory. The advanced component of this secondary radiation travels backwards in time, and some of it reaches the source on Earth. Naturally, this secondary wave is but a pale echo of the original wave, but a myriad of such pale echoes from across the universe can add up to a substantial effect. Wheeler and Feynman proved that under some circumstances the advanced secondary radiation can serve to double the strength of the retarded primary wave, bringing it up to full strength, while also canceling out the advanced wave of the original source by destructive interference. At the end of the day, when all the waves and their echoes, backward and forward in time, are totted up, the net result is to mimic pure retarded radiation. (*About Time*, Viking, 1995.)

8.07 In 1972 Bruce Partridge attempted to detect these advanced rays by means of an ingenious experiment. He beamed a rapid succession of radio pulses into the sky on a clear night, and during the intervals between the pulses he directed this same radio signal into a nearby absorbing screen. The Wheeler-Feynman effect demands that all the radiation emitted by a source today is absorbed somewhere, and sometime in the future, by the material of the universe. It does not matter how far into the future this occurs, for the returning advanced rays are supposed to arrive back at the point of emission at exactly the same time as the primary wave is emitted, but it is crucial that all the radiated energy is eventually absorbed. Partridge reasoned that it was unlikely this absorption would be complete, whereas he knew that all the energy fed into his local absorber was in fact absorbed. This difference would manifest itself as a difference in the power which his transmitter was drawing from the radio frequency generator, and he measured this power with great accuracy to detect any variation. Despite many careful attempts, Partridge detected no trace of the effect he was seeking. It might be, of course, that the universe of the future *is* a perfect absorber, and that Wheeler and Feynman were right after all, but many people would regard the experiment as confirming that advanced effects do not exist.

8.08 Most of the writers who have considered this theory have regarded it with some suspicion. Huw Price's analysis must surely be the most thorough and tightly argued treatment. He gives us twenty-nine pages of detailed and complex logical argument, and comes to the conclusion that the Wheeler-Feynman reasoning is faulty, and that assuming radiation sources to radiate equally into the past and the future does *not* result in the advanced waves being suppressed and the retarded waves behaving in the familiar way we experience.

8.09 We can agree with Price's conclusion, but should be surprised that so much ink and effort has been expended on the matter, by him and others. Feynman and Wheeler's premise, that a source of radiation emits retarded and advanced waves equally, is completely time-symmetric, and nowhere in their reasoning do they interpose any facts that are not symmetric. But their conclusion, that the wave travelling into the future is doubled while that going into the past is neutralised, is asymmetric. Surely there is no way by which symmetric premises can generate an asymmetric conclusion; there must be a flaw in their argument, even if it is difficult putting a finger on the precise place where it breaks down.

8.10 In Price's book, he goes on to present a slightly modified version of the Wheeler-Feynman theory. Significantly this does introduce a new element, and this is time-asymmetric. Price sums up his view as follows:

> What needs to be explained is why there are large coherent sources -- processes in which huge numbers of tiny transmitters all act in unison, in effect -- and why there are not large coherent absorbers or sinks, at least in our region of the universe. Of course by now we know that the former question is the more puzzling one, for it is the abundance of large sources rather than the lack of large sinks which is statistically improbable, and associated with the fact that the universe is very far from thermodynamic equilibrium. (*ibid.*)

In other words, the asymmetry of the radiation process is due to the existence of objects such as the sun, or electric lamps and radio transmitting stations, which in turn display the present low entropy of the universe, or by the fact that, like everything else of interest today, it is still constrained by the big bang boundary conditions. We cannot disagree with Price's conclusion, but he could have arrived there by a shorter route.

8.11 Viewing the transmission of light from the newer quantum viewpoint, it is difficult to see what the fuss is all about. Suppose a photographic flashgun emits a short burst of light, which from one viewpoint we must regard as a large number of photons travelling outwards at the speed of light, rather than the radiation of waves. Because each photon obeys laws which are time-symmetric, we are asked why they all travel outwards after the flashbulb is activated, and why there is not another burst travelling inwards *before* this event, to reach the flash unit just at the moment it is fired. Suppose that, instead of a flashgun we have a bomb which explodes, scattering fragments of metal in all directions. Now (ignoring air resistance) the motion of each fragment obeys the laws of elementary physics, and its motion is reversible; viewing *in reverse* a motion picture of one such fragment, it would be seen to behave quite reasonably. Must we explain why the fragments all move outwards, and why there is not also an inward rush of fragments before the explosion, to assemble the bomb just prior to its subsequent dispersal? Of course not! The bomb clearly provides an example of dissipation; before the explosion it contains a high concentration of low-entropy energy, which becomes widely dispersed after the explosion, an archetypical example of an irreversible process, and the photons radiated by the flashgun illustrate the same principle. The manufacture of such a bomb, or the charging of a flashgun, is not something nature could accomplish without a highly complicated process, which according to the picture we are advocating, links it right back to the low entropy of the big bang boundary conditions. There is no essential difference between the dispersal of the bomb fragments and of the flashgun's photons.

8.12 Other types of wave motion may be explained by different arguments, but all involve a process of dissipation, and a consequent gain in entropy, which rules out the reverse process. In the case of a stone (or an otter) disturbing the surface of a pond, it is not the material of the stone which is dissipated, as was the material of the bomb, but rather its motion. Initially the energy of the moving stone arose from its unstable position on the bank, but after it tumbles into the pond the motion is gradually dispersed, firstly as water waves on the surface, and then finally as heat when the waves are damped around the shore line. It appears that the temporal asymmetry of radiation is just one example of the Second Law of Thermodynamics at work.

8.13 However, we have considered only macroscopic examples, involving large numbers of particles. At the sub-atomic level all processes (with the unimportant exception of the decay of the neutral kaon) are reversible, and yet radiation continues to play a part in their description. We have not yet explained why such radiation also is irreversible. If we consider firstly a typical experiment in which we set up some apparatus to emit *large numbers* of photons, it is clear that in doing so we are producing a low entropy situation, a situation very unlikely to arise by chance, and we do not expect reversibility. The photons can be considered as particles, and their dispersion is a simple thermodynamic phenomenon. We can also describe their behaviour using probability amplitudes without paradox, for probabilities (and quantum theory's complex amplitudes) behave in exactly the same way as large ensembles of particles.

8.14 The emission of a single photon, however, *does* present us with a problem. Considered from the particle point of view the transference of a particle from one collision to the next is a simple reversible occurrence, but the wave function which describes it appears to radiate outwards as a spherical shell, in a manner which is clearly irreversible. We have only one particle, to which we cannot apply the argument of dissipation, or appeal to the Second Law. It is here, however, that the Nodal theory comes to the rescue, and removes the paradox. That component of the wave which relates to an expanding shell of radiation, to the travelling outwards from the first node, is purely subjective. It belongs to the CWF but not the NWF. The probabilities which this conventional wave encapsulates are based on our *knowledge* of the starting point and our *ignorance* of the destination of the particle, a distinction which has no counterpart anywhere other than in our own minds. The NWF, the wave that relays information from node to node, *is* reversible. We will analyse the form taken by the NWF in Chapter 11.

8.15 The motivation behind the search for advanced waves, a search dedicated to explaining the irreversibility of radiation phenomena, has been misguided. There are four different models we can use in discussing the radiation of light, four different pictures we can hold in the mind.

(i) The classical picture, as advocated by Maxwell, sees the radiation as the interplay of electric and magnetic fields in the space between the source and the destination of the light rays.

(ii) In the alternative quantum view, the Schrodinger wave function takes the place of an electromagnetic disturbance, but is found to propagate in the same way. At the point of reception we convert these into probabilities, which exactly replace the Maxwell intensities.

(iii) From the Planck viewpoint we must regard light as the emission and absorption of a stream of photons, a view which is forced upon us by the particle properties revealed in some of our experiments.

(iv) The Nodal picture concentrates on the photons, but claims that these are no more real than the waves; they make themselves known only at the nodes where they "collide" with particles of matter, but the information transferred from node to node by the NWF matches exactly what we can imagine is carried by the moving photons of (iii) above.

8.16 Most of the discussion about advanced waves has been conducted in terms of the first of these representations, the Maxwellian radiation, but could equally well have embraced also the Schrodinger wave equation. In both cases, another set of solutions is obtained if we replace t by $-t$, giving advanced waves which seem never to be observed in practice, but we have shown above that the search for advanced waves may well have been based on an improper interpretation of these solutions. If, instead, the third picture had been adopted in the argument, the outward flow of photons from radiating bodies, it seems doubtful that advanced waves would ever have been suggested, for the situation is so like that of an exploding bomb, or the flow of water from a leaking vessel. These phenomena are readily explained by the Second Law; we do not ask why they are never observed in reverse. All the other forms of radiation, such as water ripples and sound waves, are produced by co-ordinated macroscopic events, and their directionality also is explained as thermodynamic dispersion.

8.17 The Second Law and the Nodal theory between them teach us that the search for advanced radiation has been mistaken, and absolve us from having to continue the quest.

Chapter 9: The EPR Problem

9.01 The paper published in 1935 by Einstein, Podolsky and Rosen, *Can Quantum-mechanical description of Physical Reality be considered complete?* (Physical Review 41, 777) has probably prompted more books and articles, and promoted more discussion and argument, that any other aspect of Quantum physics. We shall in this chapter attempt our own analysis of the problem, and try to clear away some misunderstandings, before showing what contribution the Nodal theory can make to its interpretation.

9.02 The authors of the original EPR paper described a thought-experiment in which an elementary particle is made to split into two equal parts, which than fly apart. Because of the law of momentum conservation, the two parts must travel with equal and opposite velocities. Now the complementarity laws of Bohr's quantum theory decree that if we know accurately the position of a particle we cannot at the same time know its momentum. Furthermore, Bohr maintained that this lack of precision was not the result of clumsy measurement; we can talk about the momentum of a particle only in relation to a particular arrangement of measuring apparatus, which could not exist alongside apparatus for measuring the position of the particle at the same time. Bohr seems to say that the momentum *has no real existence* except in relation to the apparatus used to measure it. But the EPR paper pointed out that if we measure the momentum of *one* of the pair of particles we immediately know the momentum of *the other*, and the authors maintained that this shows the latter momentum to have a real existence, whether or not it is measured. They wrote:

> If, without in any way disturbing a system, we can predict with certainty (i.e., with probability equal to unity) the value of a physical quantity, then there exists an element of physical reality corresponding to this physical quantity. ... [It follows that] quantum mechanics does not provide a complete description of physical reality.

Bohr defended the opposite point view with arguments such as the following:

> The apparent contradiction in fact discloses only an inadequacy of the customary viewpoint of natural philosophy for a rational account of physical phenomena of the type with which we are concerned in quantum mechanics. Indeed the finite interaction between object and measuring agencies, conditioned by the very existence of the quantum of action, entails ... the necessity of a final renunciation of the classical ideal of causality, and a radical revision of our attitude towards the problem of physical reality. ... From our point of view we now see that the wording of the above-mentioned criterion of physical reality proposed by Einstein, Podolsky and Rosen contains an ambiguity as regards the meaning of the expression 'without in any way disturbing a system'. Of course there is in a case like that just considered no question of a mechanical disturbance of the system under investigation during the last critical stage of the measuring procedure. But even at this stage there is essentially the question of *an influence on the very conditions which define the possible types of predictions regarding the future behaviour of the system.* Since these conditions constitute an inherent element of the description of any phenomenon to which the term 'physical reality' can be properly attached, we see that the argumentation of the mentioned authors does not justify their conclusion that quantum-mechanical description is essentially incomplete.

The two men continued to discuss their differences for several years, and had come no closer to agreement at the time of Einstein's death in 1955.

9.03 The actual thought-experiment described in the EPR paper was never actually performed. But in 1951 David Bohm shifted the discussion sideways by suggesting an alternative experiment which might be easier to perform in practice, and which also widened considerably the philosophical challenges which it posed. It is experiments of the type Bohm invented which have subsequently come to be accepted as the authoritative version. The pair of identical particles which these experiments envisage are to be either protons, whose spins must be oppositely directed in order to conserve angular momentum, or photons which, if linearly polarised, must share the same axis of polarisation. The measurements which are to be performed on the two particles are measurements of spin, which in the case of photons are easily carried out by attempting to pass them through polarising filters. If a photon is transmitted by such a filter, conventional quantum theory maintains that its axis must then become oriented with that of the filter, whereupon conservation laws decree that the other particle of the pair must at the same moment be affected in the same way, however remote it is. Two conclusions follow immediately. If the two filters have their axes *parallel* to each other, then whenever one photon is transmitted by the filter it encounters, the other photon must be transmitted also, and whenever a photon is stopped by its filter, then so must the other. If, alternatively, the filters are set up with their axes *perpendicular* to each other, then one photon of the pair must always be transmitted, and the other absorbed.

9.04 There seemed little doubt that this effect would be observed if the experiment could actually be performed, and many writers then assumed this would show the two particles were communicating with each other in some way after they have separated, or were somehow "aware" of the setting of the polarisers before they encountered them, for how otherwise could they display the degree of co-operation observed? Many books written by physicists and mathematicians during the last thirty years describe this effect, and several of their authors declare, erroneously as we shall show, that the results prove the existence of such *non-local* influences. The EPR effect does present us with some strange contradictions, but they are not displayed in the simple case we described above, where the two polarisers have their axes parallel or perpendicular. Paul Davies, dealing with the case in which the axes of the two polarising filters are *parallel*, writes:

> The truly mind-boggling implications of [the experiment] are apparent if we use two parallely orientated polarizers, one of which is stationed in the path of each of the two correlated photons. Because the polarizations are forced to be parallel, whatever we measure for the photon polarization of one we are obliged to find the same for the other, but as there are only really two polarization states that are measurable (i.e. parallel and perpendicular) the 'yes-no' decision of one polarizer must be identical to that of the other. ... The mystery is, how does the second polariser *know* that the first one has let a photon pass, so that it too may do the same? The experiments could be carried out simultaneously, in which case we can be sure, on the basis of the theory of relativity, that no message can travel faster than the photons themselves between the polarizers to say 'let this one pass'. In fact by stationing the polarizers at

different distances from the decaying atom we could arrange for either experiment to be performed before the other, thereby ruling out any question of one polarizer signalling the other, or causing it to accept or reject a photon. (*Other Worlds*, Penguin, 1980)

9.05 There is no need for Davies' astonishment. We may agree that information could not be transmitted from one photon to the other fast enough to ensure their cooperation, but the two photons *themselves* can carry information, and may well have "agreed" a common policy before they separated. We will show a simple method which photons could employ to meet these requirements without having to pass any information to each other after the moment when they separate. Suppose their (common) spin axis is chosen randomly within the range 0° to 180°, and that a photon always passes through a polariser if its spin axis makes an angle of less than 45° with the optical axis of the polariser, and is always stopped by the polariser otherwise.

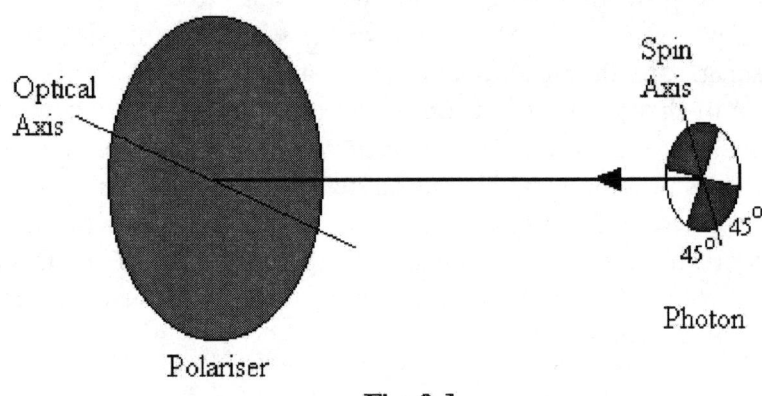

Fig. 9-1

In the diagram, we are supposing that a photon is transmitted by a polariser whenever the axis of the polariser lies within the red sectors, but is absorbed if the axis lies in the white sectors. It is easy to see that, whenever a random photon meets a polariser, the probability of transmission is 50%, as it should be, and in the experiment we are considering, with the optical axes of the two polarisers parallel, the two photons are always either both transmitted or both absorbed. It is equally easy to see that, in the alternative case where the polariser axes are *perpendicular* to each other, one photon must be transmitted and the other is not. We are not suggesting that this is indeed the explanation of how the EPR results are obtained, but the picture does show that the results quoted by Davies are not as remarkable as he imagines. The true EPR paradox is much more subtle.

9.06 So far we have considered only cases in which the two polarising filters have their axes parallel or perpendicular. If the angle between the axes has any other value, say $q°$, then quantum theory predicts, and no-one doubts, that the results of our experiment would be as follows: the probability that both photons are transmitted or both are absorbed will be $\cos^2 q$, while the probability that one is transmitted and the other absorbed will be $\sin^2 q$. It was in 1964 that John Bell considered in detail these probabilities, and discovered a remarkable result, which has come to be known as "Bell's Inequality". He envisaged a series of experiments like the one described above, except that we are able to rotate the polarisers so that their axes may be set at *any* angle. Then, making a number of very reasonable assumptions, he showed that the results must obey this inequality, unless they were influenced by some strange effect such as the instantaneous transference of information from one photon to the other. He showed further that the results predicted by quantum theory did *not* obey the inequality, which would seem to confirm that such an extraordinary effect must be at work. The special cases considered earlier, with q equal to $0°$ or $90°$, do not display these effects; they become evident only if we use a number of carefully chosen different values for q.

9.07 The final episode in this saga was written in the 1980's, when improved laboratory techniques made it possible for the first time to perform some of the experiments which previously had existed only in imagination. As expected, the results do confirm the standard formalism of quantum theory; Bell's inequality is indeed sometimes infringed by the results of these experiments. So what is Bell's theorem. and what conclusions can we draw from the experiments which violate it? A mathematical statement of the inequality can be found in many textbooks, but the difficulty of fully understanding it and its implications is increased by the fact that several different forms exist, which at first sight are not the same. We shall not present any of these forms here, but instead simply show the contradiction that is revealed by one simple series of experiments using the apparatus we have described above.

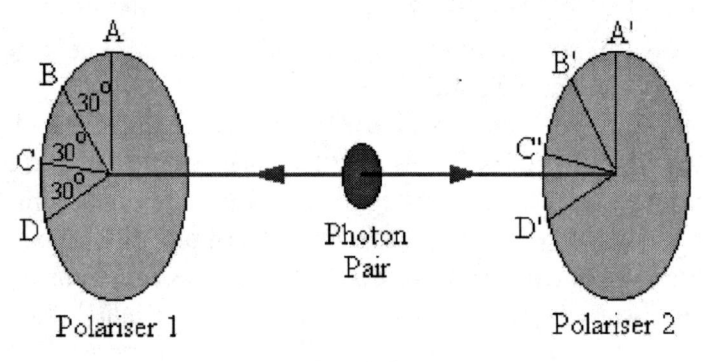

Fig. 9-2

9.08 Suppose the optical axis of the polariser on the left can take any of the positions A, B, C or D, these being at 30 degrees to each other as shown, and those on the right, A', B', C' and D', are respectively parallel to them. For our first experimental run, set the axis of the left detector to direction A and the right one to B'. The angle between them is 30 degrees and so, as the above formula suggests, the probability of agreement is $\cos^2 30°$, or 3/4; with a sufficiently long series of readings, about 3/4 of the results at A, whether "yes" or "no", will be the same as those at B'. Only 1/4 will be different, which means that the list of A results would need only 1/4 of them changing to give the B' results.

If instead of setting the left polariser to A it had been at C, while the right hand polariser remained at B', this in itself could not affect the run of results at B', which would be the same as before; we are assuming no *non-local* effects occur. And as the angle between the detectors is still 30 degrees, we would again find only 1/4 of the right hand results to differ from the left.

Thirdly, turn the right polariser to D', leaving the left one at C. Once again only 1/4 of the D' results will differ from the C results.

9.09 To summarise, only 1/4 of the A results differ from the B' results, only 1/4 of the B' results differ from the C results, and only 1/4 of the C results differ from those at D'. It follows that no more than 3/4 of the A results can differ from those at D'. But this is not what we observe; D' is perpendicular to A, and so *all* its results should differ. This contradiction is, in effect, an example of the sort displayed by infringements of Bell's Inequality.

9.10 What exactly is the significance of this infringement? The usual analysis goes something like this. Bell made two assumptions in deriving his inequality. The first was that no causal influences could be transmitted at a speed greater than that of light; we call this the *locality* assumption. Bell's second assumption was that the quantities we measure in quantum mechanics, in this case the orientation of the photons' axes, really do exist whether or not we attempt to make a measurement, the so-called *reality* assumption. So the fact that Bell's Inequality is broken, so the argument goes, implies that one or both of these assumptions, *locality* or *reality*, must be false. It is because the abandonment of either of these "obvious" precepts is so strongly opposed to intuition, that so much effort has been devoted to testing the EPR results experimentally, and to analysing the real implications of the results.

9.11 But there is a damaging flaw in the argument we presented above. We omitted an implicit stage in the discussion. We wrote, in effect, "With the polarisers at A and B' the probability that the readings differ is 1/4. So after a long run of trials, about 1/4 of the B' results will differ from the A results. Similarly about 1/4 of the C results will differ from the B' results." But we omitted to say, "Therefore *no more than 1/2 of the A results can differ from the C results.*" Now this is meaningless, for there is a fundamental impossibility in obtaining simultaneously a run of A results and a run of C results. We cannot combine into one sample space the sample spaces corresponding to the A and B' settings and the C and B' settings. The former has possible values as follows:

```
        A transmits     B' transmits    (prob =  3/8)
        A transmits     B' absorbs      (prob =  1/8)
        A absorbs       B' transmits    (prob =  1/8)
        A absorbs       B' absorbs      (prob =  3/8)
```
The latter has possible values as follows:
```
        C transmits     B' transmits    (prob =  3/8)
        C transmits     B' absorbs      (prob =  1/8)
        C absorbs       B' transmits    (prob =  1/8)
        C absorbs       B' absorbs      (prob =  3/8)
```
If we could design an experiment with all eight of the above results

as alternatives, we could combine them into a single sample space, and find the probabilities of each of the eight possible outcomes. But A and C are not possible alternatives. No experiment could determine both the A results and those at C, for any determination of the photon's state by the A polarising filter will destroy the possibility of determining how it would have responded to the C filter. No combined sample space is possible, and the results at A are not related in any rational way with the results (in a different experiment) at C.

9.12 We can express this differently. We claimed above that "if instead of setting the left polariser to A it had been at C, with the right hand polariser remaining at B', the run of results at B' would be the same as before". This is a meaningless conterfactual. Philosophers have much difficulty in explaining the real meaning of counterfactuals, and the standard work on the subject, David Lewis' book *Counterfactuals* (1973), leaves many questions unanswered, as I try to show in a publication on the internet (*Conditionals and Counterfactuals*, 1997). I believe any counterfactual in which the antecedent is not only false, but is essentially *impossible*, to be meaningless. This treatment of the EPR paradox provides a good example of such a meaningless statement. If the left-hand polariser is at A then it is fundamentally impossible to determine what the B' results would have been if the left polariser had *not* been at A. Under these circumstances I maintain that no meaning can be attached to the statement that "the run of results at B' would be the same as before", and once again we find that no conclusion can be drawn concerning the relationship of the A results to the C results.

9.13 So far as we are aware, the only writer who has realised fully what the observed infringements of Bell's inequality do and do not imply is Thomas Brody (*The Philosophy behind Physics*, Springer-Verlag, 1993). He shows that Bell implicitly made a *third* assumption in deriving his inequality, in addition to assumptions of realism and locality. He calls this the *Joint Measurability Assumption*. He writes:

> The joint measurability assumption refers to the possibility of measuring two (or more) physical quantities without mutual interference; this last expression is to be understood in the sense that neither measurement affects the value obtained by the other... In the case of spin projections, the JMA is violated.

He is referring here, of course, to the fact that a photon having passed through a polarising filter is supposed to be put into a new state, so that its original axis is no longer measurable. This is saying, in effect, what our own two statements above have said, that two measurements on the same photon cannot figure in the same sample space, or alternatively, that it is meaningless to discuss what result such an alternative measurement would have given *if* it could have been performed.

9.14 So the failure of Bell's inequality in describing the results of EPR experiments tells us only that there must be an error in *one* (or more) of Bell's three assumptions, *reality, locality* or *joint measurability*. As the last of these is false, we can make no deduction concerning the truth of the other two. The questions of the *reality* of measurable quantities, or the *locality* of quantum causality or influences, remain open, and we must look elsewhere to resolve them.

9.15 Although we are not forced to abandon either the principle of *locality* or of *reality* by the EPR experiments, because Brody presents us with a third alternative, it does seem that the *probability* which we associate with a particular photon being transmitted by a particular filter *is* inflenced in a non-local way by the orientation of the other filter, and by whether or not the twin photon is transmitted. Perhaps the two photons are governed by just one wave function rather than two, but some influence does appear to be transferred at superluminal speed. This can not be described as a *causal* influence, and it is insufficient to transfer information, and so infringe Special Relativity. The non-local influence on probabilities has no observable effect on any single pair of particles, for we cannot deduce the probability of an event merely by observing one instance of it; only when we observe a large number of particles do the effects of probability become evident. Indeed, no finite run of successive experiments, however long, can *prove* that the probabilities are influenced non-locally, any more than a run of "heads" when a coin is tossed repeatedly can prove that the coin is biased. But there is no doubt this influence exists, and it remains a strange phenomenon, with no parallel in everyday life.

9.16 We feel some discomfort when contemplating such *non-local* effects, even if they apply only to the probabilities of events happening rather than the occurrence of individual events themselves, but this discomfort is relieved to some extent by the Nodal theory. The most unpalatable aspect of the picture we painted is of the photon which has not yet encountered its polariser suddenly changing its spin axis because the other photon has encountered *its* polariser. Nodal theory teaches us, of course, that neither photon actually exists at this time, and all that is demanded is that, when the (imagined) photons reach their next nodes, the information fed into each node takes account of the orientation of the two polarisers, a picture which seems a little more credible.

9.17 Furthermore, as we have already shown, such *non-local* effects are an essential requirement of the nodal theory; indeed any particle in "moving" from one node to the next must "feel out" the whole of the surrounding region of space-time in order to "choose" its next node, and to determine what information is to be transferred to it. The present discussion extends this concept to cover the rather special circumstances encountered when a pair of particles become entangled, as they are in EPR experiments, and the behaviour of *each* is influenced by the region around *both* of them. So accepting the Nodal hypothesis means that our credulity is stretched no further by the EPR results than by other more common observations.

Chapter 10: The Special Theory of Relativity

10.01 A proper understanding of the Nodal Theory requires some familiarity with the Special Theory of Relativity, and this chapter presents a very brief account of Einstein's theory. Only those topics which are directly involved in the Nodal theory are covered here, and readers who would like to pursue the subject more deeply are advised to consult one or more of the many books available. Einstein's own little handbook, *Relativity*, translated by Robert W. Lawson (Methuen & Co.), is highly recommended.

10.02 Towards the end of the nineteenth century there was much discussion on whether it is meaningful to talk about absolute states of *rest* and *motion*, or whether the only sort of motion that makes sense is *relative* motion. In everyday life we think of trains and aircraft as moving, and houses and mountains as stationary, but when challenged we are ready to acknowledge that nothing on the earth's surface is really motionless. Our houses are moving at several hundred miles per hour because of the earth's rotation, and the earth itself moves at almost 70,000 mph in its annual journey around the sun. The sun has its own motion within the galaxy, and the galaxy also moves within the group. Is all motion relative? Can we not discuss the motion of something in relation to space itself?

10.03 Numerous methods exist for measuring the speed of light in empty space, and to a high degree of accuracy these all give the same result, which is close to 186,000 miles per second. But what does this mean? Is this an absolute speed or is it relative to something else, and if so, what? There are only three possibilities. Those who believe that all motion is relative might tell us either (i) that the speed of light always has the same value in relation to the *source* of the light waves, or (ii) that this constant speed is relative to the *observer* of the light, or the object on which it falls. On the other hand those who claim that *absolute* motion is meaningful need not specify to what the motion of light is relative; they tell us (iii) that light moves at 186,000 miles per second relative to *space*. As the nineteenth century physicists wrestled with this question they were baffled to find that each one of these three possibilities led to contradictions. Let us now examine each of them.

10.04 The first theory, that light always travels at the same speed relative to its *source*, is easily dismissed. Astronomers have discovered very many *double stars*, pairs of stars which revolve around each other, often taking no more than a few days to complete each revolution. So for part of the journey a star can be moving towards the earth, and for another part away from the earth. The light from some of these stars has travelled for many years before reaching us, and if its speed were related to that of the star which emitted it, we would observe some strange effects indeed. The light originating from the star as it moves towards the earth would eventually *overtake* that emitted a few days previously as it moved away from us; we would sometimes see the star in several positions at the same time, and sometimes we would not see it at all. Such weird effects are never observed, so we can rule out the first of our three suggestions.

10.05 The second possibility, that the speed of light is constant relative to its observer, also seemed to pose contradictions. Suppose you are at rest relative to a nearby beacon, and you are measuring the speed of the light which reaches you from the beacon. Then, while continuing to measure the light's speed, you set off in a car and travel away from the light source at 60 mph. After one minute you will be one mile further from the beacon, and its light will be taking an additional 1/186,000 second to reach you. So during the time you were travelling, the light *must* have been passing you at less than its usual speed; this second possibility also seems to be wrong.

10.06 It was the third suggestion that was favoured by most scientists in the 1880's and 90's. They believed that light waves had much in common with sound waves, and just as sound consisted of disturbances in the air, and travelled at a constant speed relative to the air, so light was supposed to consist of disturbances in some kind of *ether*, and the speed of light was constant relative to this ether. There was much debate on whether the ether around solid bodies such as the earth and the moon was dragged around with them as they moved, or whether it just flowed through them undisturbed. The first of these proposals was ruled out when it was found that, as the moon moved over a distant star during an *occultation*, there was no disturbance to the apparent position of the star right up to the moment of its disappearance, for this implied that the ether near the moon's surface was not being moved along by the moon's motion. The second proposal, that the ether passed through a moving body just as if it were not there, could be tested by experiments on the earth's surface, for then there would be an ether "wind" blowing through the earth, just as a traveller in an open vehicle experiences a wind on his face, and this would result in light having different velocities in different directions. The most famous and the most successful of the experiments to detect the ether wind was carried out by Albert Michelson and Edward Morley in 1887. In effect they compared very accurately the speed of a light ray traversing their apparatus in a North-South direction with one moving in an East-West direction. They then turned the whole experiment through $90°$, expecting that this would change the timings if an ether wind were passing through the labaratory, but no change at all was detected. Just in case they had, by coincidence, performed the experiment on the very day that the earth happened to be stationary in relation to the ether, they repeated it six months later. Once again no trace of ether wind was found. We seem to have ruled out all three of our possible assumptions.

10.07 To overcome such complete deadlock demanded extreme measures and clear thinking, and much credit is due to H. A. Lorentz (1853 - 1928) for laying the foundations which eventually showed us the way out of the impasse. Lorentz attempted to reconcile the second and third of the above suggestions. He retained a belief in the universal ether, but believed it might still be possible for every observer to find the same value for the speed of light, whatever his state of motion, if this motion itself affected the process of measurement in some way. He proposed that the equipment with which the moving observer measured the light's speed might *contract* slightly along the direction of his motion. But this contraction would not be noticed by the observer himself, for if he tried to measure the length of the apparatus he would need some sort of ruler to do so, and this ruler also would contract in the same ratio. Even the observer, and the width of his eyeball, would contract to this same extent, and so as far as he is concerned, no contraction occurs. But to the stationary observer, if he had some clever way of accurately measuring the moving ruler, the contraction would be apparent. In order to get the right answers, Lorentz maintained that not only are the moving observer's measurements of *length* affected, but his measurement of *time intervals* would also be changed, and the two inaccuracies would combine to give just the right value for the speed of light, measured in any direction. Lorentz declared that whenever a body is moving through the ether, then all lengths measured along the direction of its motion suffer this *Lorentz contraction,* and anything which can measure time on the moving body experiences this *time dilation.* These effects are not quite as unreasonable as may at first be thought, for matter was known to contain electrically charged particles, and it was possible that these could be affected by the ether wind when any body was in motion.

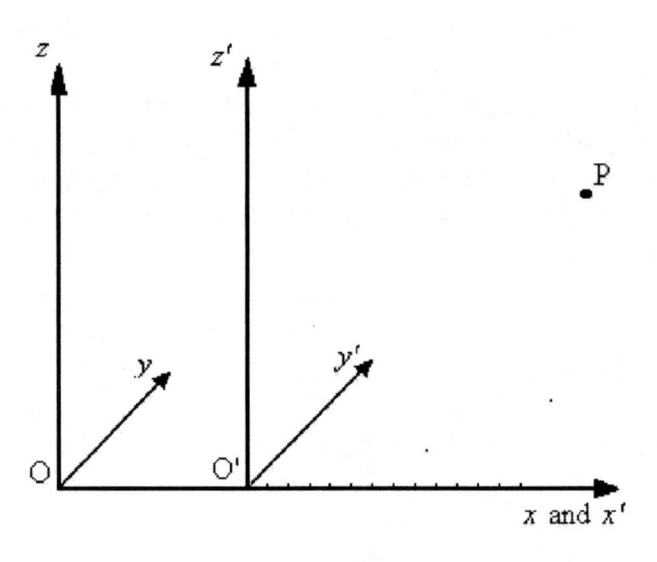

Fig. 10-1

10.08 The Lorentz effect is illustrated in the above figure. Suppose O is an observer who we consider to be stationary, and Ox, Oy and Oz are a set of co-ordinate axes he sets up in order to specify the position of any point in space, such as P, which we suppose to be fixed at the point (x, y, z). O' is a second observer who is moving along the x-axis with speed v, and O'x', O'y' and O'z' are the axes which he carries along with him. It is clear that the two observers will agree on the y and z values of the point P, for there is no relative movement in these directions, but the x'-value of P, as observed by O', will be constantly changing because of the relative motion. If t is the elapsed time since O and O' coincided, then (before Lorentz had proposed his new theory) we would have expected the following relationships to apply :

$$x' = x - vt$$
$$y' = y$$
$$z' = z$$
$$t' = t$$

Formulae 10-1

The fourth of these equations is inserted because it seems obvious that the two observers will agree on what time it is. But now Lorentz teaches us that:

$$x' = \frac{x - vt}{\sqrt{(1 - v^2/c^2)}}$$
$$y' = y$$
$$z' = z$$
$$t' = \frac{t - vx/c^2}{\sqrt{(1 - v^2/c^2)}}$$

Formulae 10-2

where c stands for the speed of light. These formulae show the manner in which measurements of length and time differ between two observers who are in relative motion to each other, and these differences do remove the paradox concerning the speed of light.

10.09 To see how the Lorentz formulae describe the contraction of a measuring rod when viewed by a moving observer, suppose the stationary observer is holding the rod parallel to his x-axis, and that the moving observer attempts to measure its length as he passes. To the stationary observer the length of the rod is the difference between the x-co-ordinate of its two ends, namely $x_2 - x_1$. But the length of the rod as seen by the moving observer would be $x_2' - x_1'$. Using the above formulae it is easily shown that the latter result is less than the former. The length as viewed from the moving reference frame is less than that in the stationary frame, assuming, of course, that each observer looks at the two ends of the rod at what he considers to be the same instant of time (and makes due allowance for the finite speed of the light rays by which he sees them).

10.10 This is such an unexpected result that one may ask why it was not discovered experimentally long ago. The reason is simply that all the velocities with which we deal in everyday life are much less than c, the velocity of light, and the formulae show that the contraction effect is then very small indeed. If v is 5 miles per second, about the speed of an orbiting earth satellite, the contraction is less than one part in a thousand million; but it becomes more significant as v approaches the value of c, and theoretically an object's length would vanish completely if it could reach the speed of light.

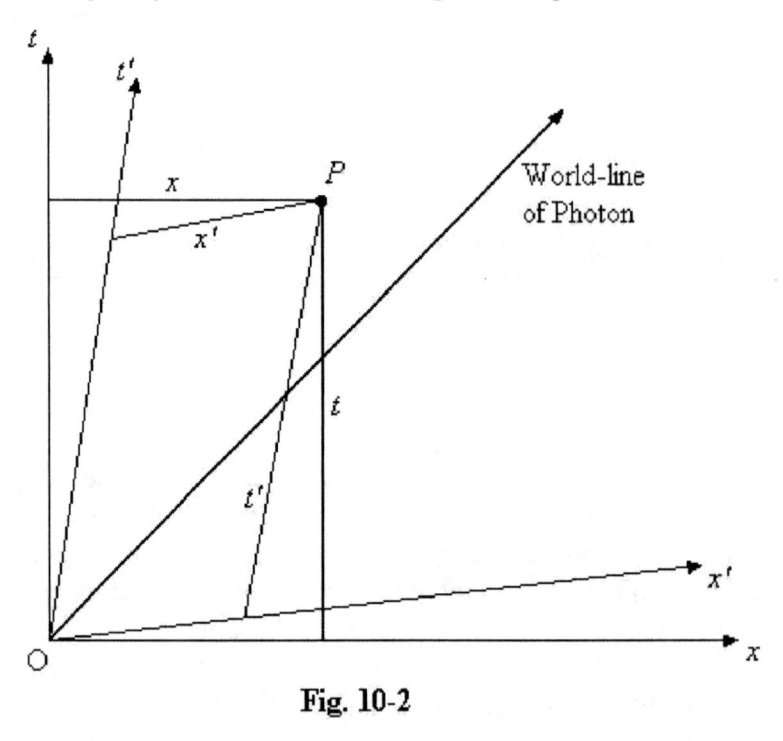

Fig. 10-2

10.11 It is instructive to view the relationship between the two reference frames on a diagram which shows the *time* co-ordinate as well as the *spatial* co-ordinates, as we did in Chapter 2. We can visualise no more than three axes, and so must abandon one or more of the space co-ordinates, and in fact on this diagram we ignore both the y and z co-ordinates, but this does still allow us to see the important relationships between the x's and the t's. The t axis represents the world-line of the stationary observer, and the t' axis that of an observer moving with speed v along the x-axis. The scale of the t-axis is chosen so that the world-line of a photon or a flash of light makes an angle of 45° with the x-axis. The Lorentz equations show us that Ox', the x-axis for the moving observer, no longer coincides with Ox, the stationary observer's x-axis, but is inclined at an angle as shown. We have represented an event P, and draw lines to indicate its co-ordinates in both systems. The diagram also shows clearly a fact we stressed in Chapter 2, that the notion of *simultaneity* is purely subjective. The x-axis represents all those events which a stationary observer at O regards as simultaneous, for every point on the x-axis has a t-value of 0. But to the moving observer at O, it is the x' axis which represents the events for which $t' = 0$, and so these are the events which this observer regards as simultaneous.

10.12 But our troubles are not yet over, for several other physical formulae are not transferred correctly by the Lorentz transformations, including all the well-known formulae of mechanics. And there was something unsatisfactory about Lorentz's theory. It claims that there is a state of absolute rest, but that nature goes to great lengths to hide it from us. This contraction of measuring rods and slowing down of clocks is cleverly contrived to prevent us ever knowing whether something is truly motionless, or whether a relative speed is really the true absolute speed. Such a conspiracy would be difficult to explain or believe.

10.13 It was at this point that Einstein came onto the scene, and he tackled the problems in a much more thoroughgoing way than anyone before him, and took as his *Principle of Relativity* the following doctrine:

> If K and K' are any two co-ordinate systems, and K' moves relative to K at uniform speed and without rotation, then *all* the general laws of physics take exactly the same form in K' as in K.

Basing his calculations only on this one principle Einstein resolved all the outstanding problems with his Special Theory of Relativity. The principle does rule out the possibility of an absolute standard of rest, for this would single out one particular co-ordinate system. But Einstein showed that the Lorentz effects did not, after all, require the existence of the ether, and he made them seem much more plausible by showing that the contraction and time dilation were not real physical changes, but just appeared naturally because of the different systems of measurement used by the two observers.

10.14 Perhaps the most noteworthy of Einstein's conclusions concerns the notions of *mass* and *energy*. He show us that the mass of an object is not a constant, as had always previously been thought, but must be considered to increase as the object's speed increases. We shall use the symbol m_0 to represent the body's *rest-mass*, and we find that the actual mass at speed v is given by:

$$m = \frac{m_0}{\sqrt{(1 - v^2/c^2)}} \qquad \text{Formulae 10-3}$$

Einstein teaches us that *mass* and *energy* are really manifestations of the same characteristic of a body, and that a body must be considered to contain energy even when it is not moving. The *rest-energy* of a body is found to be

$$E_0 = m_0 c^2 \qquad \text{Formula 10-4}$$

and the total energy of a moving body, including its kinetic energy, is given by

$$E = \frac{E_0}{\sqrt{(1 - v^2/c^2)}} \qquad \text{Formulae 10-5}$$

This leads us to the best-known of all the formulae of Special Relativity,

$$E = mc^2 \qquad \text{Formula 10-6}$$

where m is the total mass of a body as given by formula 10-3 above. Calculations show that the classical textbook formula for the kinetic energy of a moving object, $K = mv^2/2$, is not accurate at high speeds, whether or not the value of m takes account of Einstein's formula 10-3. We can find the true value by subtracting E_0 from E, and for values of v not too near to c the result can be expressed approximately as follows:

$$K = \tfrac{1}{2} m_o v^2 + 3 m_o v^4 / 8 c^2 \ldots \qquad \text{Formulae 10-7}$$

The only other result required in our treatment of the Nodal Theory is the formula for the *momentum* of a body. This is given in relativity theory, as it is in classical theory, by

$$p = mv \qquad \text{Formula 10-8}$$

but care must be taken to use the correct value of m, and not m_o.

10.15 The writer regrets it is has been necessary to treat Einstein's theory so cursorily, but a fuller treatment would have been out of place in a book on Quantum Theory. He does hope that readers who have not studied this beautiful branch of physics will explore it further in other books or web pages.

Chapter 11: The Nodal Wave Function

11.01 We shall now look more formally at nature's own wave function, the NWF (nodal wave function), and examine its relationship with the wave function used by quantum scientists in their everyday calculations, the CWF (conventional wave function). We shall consider as an illustration a very simple situation, that of a particle such as an electron which is "moving" from one node to the next in free space, a situation which we like to think of as a uniform motion in a straight line. We may have some definite information about the electron's past history, and so have partial knowledge of the position of the first node and of the particle's energy and momentum. In such a situation the CWF used by quantum physicists to represent their knowledge of the particle may take the form of a wave-packet, whose size is a measure of their uncertainty concerning the particle's position. We briefly discussed such a wave packet in Chapter 4, (see Fig. 4-1), and discovered that because it is limited in extent, it must contain not just one wavelength but a whole range of values. We know that wavelength is related to the momentum of a particle by the de Broglie relationship $\lambda = h/p$, where $p = mv$ is the momentum, and so the spread of possible wavelengths reflects our uncertainty of this momentum.

11.02 This picture of an electron's wave packet does bear some relationship to reality. If we imagine (falsely) that the particle travels in a straight line at constant speed, its position at any instant should lie within the limits of the moving wave, and the true wavelength of the particle will lie within the limits set by the uncertainty principle. But in several respects the picture is false, for the *spread* of wavelengths, and the associated limited length of the wave packet are due solely to our own limited knowledge of a particle whose future history we do not know. The positions in space-time of the two nodes are perfectly definite, so there is no intrinsic uncertainty in the momentum, but the movement of the wave packet from node to node does not represent anything in the real world, for the electron does not really make the journey. Because the waves represent the electron's *probability amplitude*, however, the wave function does indicate to us the probability, based on our present information, that the particle's next node, *if it occurs at the moment represented in the picture*, will lie within the packet. We can imagine the packet moving to the right in our diagram, preserving the same wave structure as it moves, but here again we are in error. Mathematical analysis shows that the average speed of the packet is indeed *v*, as we would expect, but the waves themselves, as distinct from the packet, move at a different speed. In fact the waves are moving *through* the packet with speed c^2/v, where *c* is the speed of light. (For a simple derivation of this result see *Modern Physics*, by Gautreau and Savin, pub. McGraw-Hill, p.80.) This speed is usually far in excess of *c* itself; if *v* is one-tenth of *c* then c^2/v is ten times *c*. We cannot observe or measure this speed, and it does not represent anything very real. So far as we are concerned it has little more significance than the speed of a shadow moving across the floor, a speed that may often exceed that of the object casting the shadow. This superluminal speed of the de Broglie wave need give no more cause for concern than the speed of a shadow.

11.03 The conventional wave packet thus contains only imperfect information about the particle between its two nodes, but it does contain all the knowledge we can have of it until we make our next observation. Furthermore quantum theory ensures that, if we repeat an experiment many times, the probability density (based on our incomplete information) of this next node lying at any particular point within the packet, is given by $|\Psi|^2$, the squared modulus of the complex wave function at that point. So while some features of this wave, the CWF of quantum mechanics, do codify information which nature uses to connect one node with another, other features are essentially subjective, and are related only to the limited knowledge which a particular person happens to have of the particle's history. It must also be time-asymmetric, for as the text books show, although the wave packet moves in a manner which we can associate with our picture of the moving particle, its size increases as it progresses, because of our uncertainty in the value of *v*. The real elements in such a wave, the elements which it shares with the NWF, must comprise only those features which, firstly, are independent of any individual's knowledge or lack of knowledge, and secondly, are time-symmetric.

11.04 Much of the formalism of quantum mechanics can be derived from one simple hypothesis. Let us suppose a particle of rest-mass m_o has a node whose position in space-time is A, and whose next node occurs at B, in a region free from other neighbouring particles. We shall represent only two dimensions on our diagram, one dimension of space, x, and the time dimension t. We shall simplify the picture further by viewing the situation using the reference frame in which the particle is *at rest*, i.e. for which $x = 0$ at both A and B.

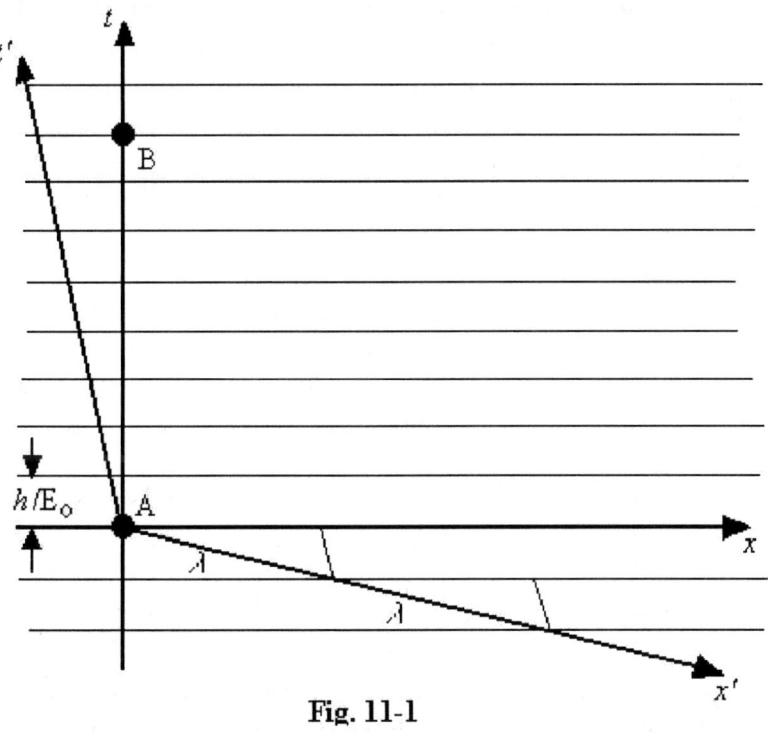

Fig. 11-1

The assumption we make is that the region of space-time surrounding such a stationary particle is filled with a uniform wave whose frequency is given by E_o/h, where $E_o = m_o c^2$, the particle's rest energy, and h is Planck's constant. The equation of the wave is therefore:

$$\Psi(x, t) = a \exp(-2\pi i E_o t / h) \qquad \textbf{Formula 11-1}$$

and we are proposing that this wave is, in fact, the particle's NWF.

- 137 -

As explained in Chapter 3, this function represents a complex number, and it can be visualised as a two-dimensional vector of length *a*, rotating at high speed. The function contains only one variable, *t*, and makes no reference to the *x, y* or *z* values of space, and so it has the same value throughout all space (in this reference frame); the rotation we imagined represents the changing direction of the vector at different times *t*, a direction that at any instant is everywhere identical. The whole of space at each instant of time is represented in our diagram by a line parallel to the *x*-axis, and we have drawn lines corresponding to those times at which the phase of our wave function happens to be zero. The frequency E_o/h depends only upon the mass of the particle, and is unrelated to any particular observer. The frequencies associated with all the common atomic particles have high values. For example, the frequency we must associate with an electron at rest is about 10^{20} (one hundred million million million) per second.

11.05 It is not possible to observe these waves, but we can gain some insight into their behaviour if we consider how they would appear to an observer moving relative to such a particle, (or what amounts to the same thing, to an observer at rest viewing a moving particle). Let us suppose the observer is moving with speed *v* along the negative *x*-axis. The *t'* axis on our diagram shows the observer's world-line, and because Special Relativity teaches us that such an observer will have a different "now" plane from that of the particle, we draw his *x'* axis also. Notice, as always, that we must not attribute any *change* or *motion* to this space-time diagram. But even if the diagram itself cannot change, we might be allowed the luxury of imagining ourselves to move steadily along the *t'* axis, to find out what we would observe, with our illusory belief in a "moving" time. At any moment of time, the events which are happening "now" lie on a line parallel to our x' axis, and as this line moves upwards on the diagram we see its intersections with the electron's waves moving rapidly away from us, to the right on the diagram. Quite simple calculations, making use of the transformations of Special Relativity (see Appendix 1), show that the wavelength along the x' axis is given by *h/mv*, which is of course just λ, the wavelength as given by de Broglie's formula. Indeed de Broglie used a similar argument in the

1920's to derive his formula, but without realising the significance of the premise from which he started. As shown in the Appendix, these calculations gives us the standard wave equation for a free particle in quantum theory:

$$\Psi = a\ \exp(2\pi i\,(px - Et)/h) \qquad \textbf{Formula 11-2}$$

where p and E refer to the momentum and energy in the observer's reference frame.

11.06 It may be thought that the set of parallel waves described by Formula 11-1 is no more than a mathematical artefact to help us calculate the value of λ for different observers. But as its layout and spacing depends only upon the mass of the particle, and is independent of its velocity with respect to the observer, it is reasonable to suppose that these waves do indeed have a real physical existence. A set of waves similar to those illustrated is associated with each segment of the world-line of a particle, i.e. between each consecutive pair of nodes. This wave is independent of the observer or his velocity, as it must be if it represents the objective NWF of the particle. It contains sufficient information to determine the *direction* in which the particle is moving (for the particle's assumed world-line is orthogonal to the plane of the waves), its rest mass (given by E_o, which determines the frequency), and at each moment of time, the phase value which is needed to specify its behaviour in interference experiments.

11.07 A clearer picture of what the observer can or cannot see is often presented if we draw our diagram in the observer's rather than the particle's reference frame. Here the *x-t* axes are the observer's, and the *x'-t'* axes are the particle's. We again show the parallel lines corresponding to the particle's NWF, and we show also the "wave group" or "wave packet" with which a particular observer might describe the particle's motion. You are asked to imagine the *x*-axis moving steadily upwards to represent the (apparent) progression of

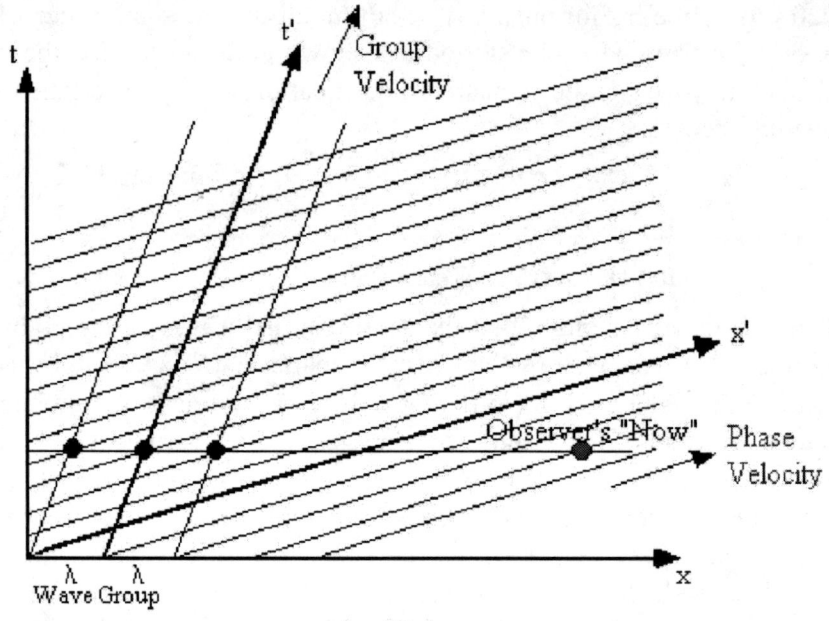

Fig. 11-2

the observer's time. The wave packet will be seen to move along with the "group velocity" v, and the NWF waves to move rapidly *through* the packet, with the superluminal "phase velocity" c^2/v, as the standard wave theory of quantum mechanics tells us they must. As explained above, we cannot in practice observe these waves, but by indirect means, for example in interference experiments, we can measure the wavelength, λ, giving the correct value for the particle's momentum (h/λ) in the observer's reference frame.

11.08 We are beginning to see which features of our conventional wave form are objectively real, and which parts are subjective. The NWF, associated throughout space and time with this segment of the particle's history, can be believed really to exist, and its form is independent of anything we choose to do. But the wave packet which we imagine to move along the *x*-axis is *subjective*, and its wavelength depends upon the observer's velocity relative to the particle. The size of the wave packet also is subjective, and in a different way; it is dependent not just on the observer's state of motion, but on the extent of his *knowledge* of the particle's past history. It has nothing to do with the objective state of affairs, but simply indicates the precision which our experimental arrangement allows in the measurement of the particle's past positions. And because we have access only to a restricted part, dx, of the wave form, our estimate of the speed v and the wavelength λ must be uncertain, as we have explained, and as decreed by the laws of Fourier analysis. In the diagram the width of the packet is about 2λ, which means that any determination we attempt to make within the packet for the value of λ will be in doubt by about $\lambda/2$. From this we deduce that our attempt to find the particle's momentum p will be in doubt by about $p/2$. So $dx.dp > 2\lambda.p/2$, which equals λp, or h, as we already know from Heisenberg's uncertainty principle.

11.09 So starting only with our assumption that the NWF of a particle at rest is described by Formula 11-1, we have derived several of the text-book formulae for the behaviour of a moving particle. The particle's NWF is manifest to us only when we are moving relative to that particle, and the transformation of Special Relativity ensures that our space makes a small angle with the crests of the wave function. The spacing of the intersections determines the wavelength of the waves we observe, and these appear to run along our x-axis at the expected speed as our time seems to advance. Proceeding along similar lines much of the standard quantum theory can be deduced.

11.10 So far we have considered only the NWF of an isolated particle in free space. In more complex situations the form taken by the NWF is closely related to the disposition of nodes throughout space-time, and we can picture the NWF as a superposition of the waves associated with every consecutive pair of nodes, and as permeating the whole of space-time. But it may be asked which partner of the duo, the wave or the nodes, is *responsible* for the other? Does the wave *determine* the disposition of nodes, in the same way that a country's road network determines the location of its road junctions and cross-roads? Or does it merely reflect a layout which is already decided, in the way that such a road network reflects the locations of the country's towns and villages? It seems likely that the question is really without meaning. Our usual criterion for deciding which of two events is responsible for the other is to ask which occurs first in time. Our villages must have been located before tracks were laid down to connect them, so the locations of the villages determined the layout of the tracks. But the crossroads could not exist before the roads themselves, and so their locations were determined by the layout of the roads. In each case the earlier event is responsible for the later. Such a question cannot be answered for the wave and its nodes, however, since the wave is supposed to fill the whole of space-time and so exists both before and after the events it describes. It seems that the question of dependence cannot be answered, and can be dismissed as either meaningless or, at any rate, unimportant. The whole of nature comprises a pattern of nodes distributed throughout space-time and a related network of waves which connects them. We are not to ask which of these is responsible for the other.

11.11 The NWF for a single particle, describing its "motion" from one node to the next, behaves in some respects like the CWF of traditional quantum mechanics, but with one important difference. We form the CWF from a position in which we know something about the time and position of the first node, but know little about the second one, and we can only calculate probabilities for it, using the distribution given by the squared modulus, $|\Psi|^2$, of the waveform. For the NWF, on the other hand, the two nodes are on an equal footing. It "knows" the time and location of both nodes, and its form relates equally to both; no uncertainties and no probabilities are involved. But from our point of view, it is sometimes valuable to use the NWF to give us probability information. Many of the experiments we set up, such as those involving interference, concern the preparation of a large number of particles in identical states, and from a theoretical point of view the NWF can then be used to give the expected distribution of final states. There is nothing unreal about the interference patterns produced on a photographic plate, and these are related to the superposition of NWF waves in the same way as that described by conventional quantum theory using the CWF.

11.12 The above paragraphs all refer to a particle such as an electron which can move only at speeds less than that of light. Before we end this review of the Nodal Wave Function we shall look briefly at the corresponding function for particles such as photons. The NWF of these particles cannot take the same form as that of massive particles, firstly because their rest-energy is zero, and secondly because there is no reference frame in which they at rest. Our approach is to *assume* that the Formula **A**, which we obtained in Appendix 1, still applies, and then derive the required NWF from this. The resulting formula, although it describes the waveform in terms of the reference frame in which the particle's *source* is at rest rather than the particle itself, and quotes the wavelength λ applicable only to this frame, nevertheless is frame-invariant. The NWF of such massless particles is derived in Appendix 2, and is as follows:

$$\Psi(r, t) = a \exp(2\pi i (x - ct)/\lambda) \qquad \textbf{Formula 11-4}$$

These particles display different wavelengths, energy and momenta in different frames of reference, just as do massive particles, but the *frequency* of the oscillations which they display also assumes importance in this case, and is often observable. Indeed ordinary radio waves take this form, and we generate them by means of familiar electrical circuitry designed to produce the required frequency. The formula, and the diagram, both show that the observed frequency is c/λ in the reference frame where the wavelength is λ, as we should expect.

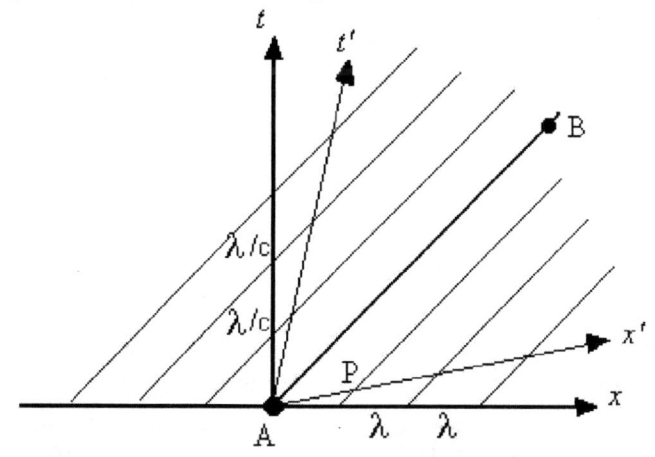

Fig. 11-3

11.13 It may seem strange at first sight that the wavefronts do not intersect the particle's trajectory, so that Ψ seems to have the same phase along the whole trajectory, but a little thought confirms that this must indeed be the case, for the relativistic time dilation effect of a particle moving at the speed of light becomes infinite; a clock moving with the particle registers the passage of no time at all between the beginning and the end of its flight. The phase of the Ψ-value is the same throughout its journey.

11.14 We should mention briefly the way the NWF of a particle can give rise to interference effects. The NWF of a particular particle can interfere only with itself, as when two adjacent parts of the wave become separated and subsequently come together, so that their Ψ values become added, or superposed. The wave of one particle does not interfere with that of another, but there are circumstances in which the NWF corresponding to one *segment* of a particle's history can become superposed on that from another segment to produce interference effects. This can happen, for example, when a particle traversing Young's apparatus collides with a low energy photon at one of the slits. We saw in Chapter 6 that the particle can still go on to produce interference effects, provided the photon's energy is low enough, so that the values carried by the wave-form in the next segment of the particle's history are almost identical with its previous values. It behaves almost as if no collision had occurred, and contributes to the interference in the same way in both cases. The interference takes place between two different segments of the particle's wave form, one having passed without interruption through the first slit, and the other resulting from the photon collision at the second one. We discussed another example of this phenomenon in Chapter 7, where one branch of a divided wave form is reflected from a mirror, and suffers a collision there, and yet can still go on to interfere with the other branch which has suffered no collision.

11.15 We shall give just one illustration of the way a diagram like that above can help us solve problems involving the NWF of a photon. Let us suppose an observer is moving along the x-axis, with speed v relative to the source A, ahead of a burst of radiation, and that it overtakes him at B. If he measures the wavelength of the wave, he will obtain a value greater than λ because of the Doppler effect. In Fig.11-3 we have drawn the x' and t' axes for the observer, and it can be seen that the measured wavelength, AP, is in fact greater than λ, as we expect. To keep the diagram simple we have centred the observer's axes on the point A where the particle is emitted, but the measured wavelength would be the same length AP wherever it had to be taken, and whatever method was used. In Appendix 2 we give the necessary mathematics to calculate the length of AP, and we obtain the result:

$$\lambda' = \lambda \frac{\sqrt{(c+v)}}{\sqrt{(c-v)}}$$

which relativity theory tells us is the correct value of the Doppler shifted wavelength for an observer receding from a source of light with wavelength λ. As in the case of subluminal particles, our postulated description of the NWF of the massless particles can lead to many of the fundamental features of quantum theory.

11.16 Throughout this chapter we have referred only to the NWF of elementary particles such as electrons or photons, but many experiments prove that interference effects can be observed also with composite bodies such as atoms and molecules. Whole atoms have been used in particle experiments for several years, and at the time of writing quite large molecules are beginning to show that they also are capable of demonstrating interference. Does this imply that such bodies have their own NWF, or can the observed effects be explained in terms of the wave functions of their individual elementary parts?

11.17 The behaviour of an unconnected group of particles differs greatly from that of a composite body because of the strong forces which bind together the component parts of such a body. A photon striking an isolated electron affects only that electron, and in particular can change its momentum, while a photon striking an electron in an atom, provided it is not sufficiently energetic to shift the electron to a higher energy level within the atom, must change the momentum of the atom as a whole. The internal forces which hold the electrons in their place around the nucleus, are sufficiently powerful to maintain the configuration of the atom, so that it rebounds as a whole. Today we picture these internal forces as being transmitted by "virtual" photons or other "messenger" particles, and these must have their own NWF's to convey information between each pair of nodes, and in particular to transmit any change of momentum to the atom as a whole. But these effects are purely internal, and none of the local wave functions can reflect the momentum of the atom as a whole. So it seems certain that every atom, every molecule, or indeed every rigid body, must have its own NWF which carries information about the body's momentum, and which under suitable conditions can be responsible for interference effects.

11.18 If a group of discrete particles happen all to be moving with the same direction and speed, each will behave as if the others were not present. When such a group encounters an obstacle, such as a diaphragm containing a narrow slit, each particle will be diffracted as if it were alone. Its wavelength is related to its momentum according to de Broglie's formula, and those with less momentum will tend to be diffracted through greater angles because of their greater wavelength. But if these particles are locked together, as are the electrons in the shells of an atom, or the

component atoms of a molecule, and this composite body survives its encounter with the diaphragm, then it is diffracted as a whole. Its momentum is obviously greater than that of each component, and so its wavelength is less, and its mean angle of diffraction will also be less. This explains in part the fact that we do not normally see interference effects in everyday life; other things being equal, the larger a body the smaller is its wavelength, and the more difficult is it to devise apparatus which will reveal it. An object such as a bullet or a pea, travelling at the sort of speed with which we are familiar, has a de Broglie wavelength many orders of magnitude smaller than any apparatus we could devise to reveal diffraction or interference effects.

11.19 One can easily imagine why the individual particles of such a composite body do not themselves give rise to interference effects. Each particle is kept in its place by the constant buffeting of the virtual photons which hold the body together, each collision being represented by a node, and causing some change in the particle's momentum. So if the body as a whole has two or more routes to a destination, as for instance in a two-slit experiment, the separate particles have no chance of preserving their coherence, just as a stream of unconnected particles produce no interference if they suffer collisions during their journey through Young's apparatus.

11.20 It is less easy to understand why the composite body is associated with a wave of its own, which under suitable conditions can itself display interference effects. The wave is in no sense a *sum* of the Nodal Wave Functions of the component particles. The essential defining characteristic of a NWF is its frequency, which we have shown is given by $m_o c^2 / h$, so that the *frequency* of the body as a whole is in fact the sum of the frequencies of its components, but this has no mathematical significance. Furthermore we have shown that the NWF of one particle never interacts with that of another. But interestingly the actual NWF of the whole body is the *product* of the component NWF's, for each NWF is given by $\exp(-2\pi i m_o c^2 t / h)$, and the exponential of a sum of exponents is the product of the exponentials of the individual exponents (or in symbols, $\exp(x+y) = \exp(x).\exp(y)$).

11.21 There is no obvious reason for this to be the case, and at present we must accept it as an unexplained fact that every composite body is associated with its own de Broglie's wave, just as is every fundamental particle. Indeed there is much which is not understood in this area. From the point of view of traditional quantum theory there are three correlations for which no explanation is known. The classical quantum physicist must accept without question that (i) every particle has a characteristic *mass*, which is additive when such a particle has component parts, (ii) that the magnitude of the effect produced when a body collides with another is determined by its momentum, and so is proportional to its mass, and (iii) that the de Broglie wavelength of a particle (as seen by an observer moving relative to the particle) is inversely proportional to its mass. Feynman understood clearly the depth of our ignorance of the true significance of *mass*. He wrote in 1985:

> ... there remains one especially unsatisfactory feature: the observed masses of the particles, *m*. There is no theory that adequately explains these numbers. We use the numbers in all our theories, but we don't understand them -- what they are, or where they come from. I believe that from a fundamental point of view, this is a very interesting and serious problem. (*QED*, p.152)

11.22 The Nodal Theory has one less problem to solve than has traditional quantum theory, for it denies the existence of the particles themselves. The "mass" of a body, and the Nodal Wave Function of such a body, do not need separate explanations, for mass *is* no more than the frequency of the wave. But nodal theory still has no explanation to offer for the remaining two questions, of why the wave frequency of a composite body can be found by adding the frequencies of its parts (or equivalently, why the NWF itself is found by multiplying NWF's), and why the magnitude of the effect of one body on another is proportional to this frequency. We know that the world-line of every composite body corresponds to a wave in space-time which we can describe exactly, but we lack totally any understanding of why this should be so. In this respect such bodies do not differ from the elementary particles of which they are composed, for again we know precisely the parameters of the NWF joining the nodes of such particles, but can give no explanation for the wave, or account for the values of its parameters.

Chapter 12: Conclusions

12.01 The Nodal Interpretation of Quantum Mechanics has nothing to say about the results of experiments. Its conclusions have no bearing on observations or measurements made in physics laboratories or on the sites of particle accelerators. The same must apply to all the other pictures of the quantum world, for if this were not so, surely experiments would have been devised which could distinguish between them, and we would have been spared nearly a century of controversy and perplexity.

12.02 So what purpose do these pictures serve, and what can be gained by adding one more to the list? They exist only to satisfy our curiosity about the *reasons* for things, to provide us with *explanations* and *understanding*. The blind formulae of quantum mechanics give correct answers in the laboratory, but the various theories attempt to provide interpretation. Galileo knew how to calculate the distance a falling body would move in a given time, but did not know why it fell. Newton gave us a partial understanding of why a body falls when he showed that the rules it follows also govern the motions of the moon and planets. Then Einstein carried this process further by showing that the *changes* in such motions, which Newton had ascribed to some intangible and unexplained gravitational force, were due solely to properties of space and time, and required the postulation of no mysterious forces acting at a distance.

12.03 Perhaps some similar all-embracing principle will one day give a generally agreed explanation for the behaviour of particles in the micro-world, but it seems unlikely that any such principle will remove entirely the perplexity which we feel when contemplating this behaviour. The quantum world is so different from anything we experience directly that it will always seem strange. The controversy which persisted through most of the twentieth century between the various interpretations of quantum phenomena was centred on the relative plausibility of the contending explanations, on their intellectual economy and tidiness, rather than on any differing predictions they made. It is into this arena of conflicting ideas that we pitch the Nodal Interpretation, believing that its simplicity, and the relatively few directions in which it challenges our credulity, make it worthy of consideration and comparison with the established theories.

12.04 While it seems unlikely that we can ever test experimentally these various interpretations, it is possible to argue against several of them on other grounds, particularly where they come into conflict with the fundamental properties of *time* which we developed in Chapter 2. The time-asymmetry of the collapse of a particle's wave function contrasts sharply with the dynamical symmetry of the particle's behaviour when seen from a mechanical point of view, and becomes an outright inconsistency when we recall that the wave and the particle are attempts to explain the *same* phenomena.

12.05 This inconsistency is again evident when we view the history of a quantum system as a static picture on the four dimensional canvas of space-time. The conventional wave-form describing the state of the system during the time interval between two events is based on the supposition that, because the first has already occurred, it is a certainty, while because the second lies in the future, and because we know the quantum world not to be deterministic, it is still undecided, and may yet be one thing or another. But, as we have stressed throughout the book, the dividing line between past and future, the moment we call "now", does not exist in the real world, and is a purely subjective phenomenon within individual brains, so there can be no intrinsic difference between future events and past ones. The future seems to us less certain than the past only because we can remember the one but not the other. *We* are less certain of the future than the past; the future itself is no less certain. The wave-form of conventional quantum mechanics is a *probability amplitude*, providing a measure of our ignorance of a future event until it occurs, and collapsing when that ignorance is replaced by certain knowledge as the result of observation. But for Nature herself there is nothing about which to be ignorant. Past and future events are all laid out in four-dimensional space-time with nothing to distinguish between them, and the *probability* of future events is an unnecessary and meaningless idea. Those features of the wave-form which appear to collapse at the moment we make an observation are wholly in the mind, with no counterpart in the real world.

12.06 The present theory, by basing itself on the picture of time which we presented in Chapter 2, forces us to ignore what our intuition tells us about the nature of past and future, the concept of "the present moment", and the apparent flowing of time. Renunciation of these beliefs may provide the chief stumbling block to a ready acceptance of Nodal Theory, but this new view of the nature of time is now becoming widely accepted, or at least widely discussed, by philosophers and physicists, including many who have no special interest in, or knowledge of, quantum mechanics. We shall quote two passages, written recently by a philosopher and a physicist respectively, which discuss the main thesis on which the Nodal theory is founded, that there is no such thing as "now" in the world outside ourselves, that the nature of future events can therefore not differ intrinsically from the past, and that the "flow" of time must be an illusion.

> ....[We shall discuss] the old philosophical controversy about the status of the distinctions between past, present and future. One side takes the view that these distinctions represent objective features of reality, the other that they rest on a subjective feature of the perspective from which we view the world. On the latter view, the notion of the present, or now, is perspectival in much the way that here is. Just as there is no absolute, perspective-independent here in the world, so there is no absolute, perspective-independent now. Obviously a view of this kind does not have to say that our ordinary use of "here" and "now" is mistaken ... Once we understand the perspectival character of these terms, we see that their ordinary use does not involve a mistaken view of the nature of reality; it doesn't presuppose that there is an objective here and an objective now. (*Time's Arrow and Archimedes' Point*, Huw Price, OUP, 1996)

> We *remember* the past, we plan the future, but we act now. The present moment is our moment of access to the universe, -- we can always change the world at this instant. But what is "now"? There is no such thing in physics; it is not even clear that 'now' can ever be described, let alone explained, in terms of physics. For example, suppose the following is tried. 'Now' is a single instant of time. The response 'which instant' yields the answer 'every instant'. Each instant of time becomes 'now' when 'it happens'. But this is going in circles. ... To counter that all time is 'now' eventually, but not all at the *same* time, is mere tautology. (*Space and Time in the Modern Universe*, P.C.W. Davies, CUP, 1977)

12.07 Let us examine again some of the problems which the quantum world presents to us, and assess the plausibility of the explanations which the Nodal theory offers. This theory, like all the others, has its own *wave function*, but it assigns a different function to represent each segment of a particle's world-line, that part of the world-line lying between each consecutive pair of nodes. The wave is independent of any observer, and is determined wholly by the characteristics of the particle and the positions in space-time of these nodes. Furthermore, it is time-symmetric; nodal theory claims to overcome the temporal contradictions which plague other theories.

12.08 The classical experiments which appear to show quantum bodies possessing simultaneously the incompatible properties of *particles* and *waves* are readily explained by the nodal theory. We claim that particles exist only at nodes, the points where they seem to come into contact; they do not move from one to the next. But as we are prevented from gaining any direct knowledge of a particle *except* where it experiences collisions, its non-existence between them can have no effect on what we observe. We can consider the *wave* to move from node to node, but those features of the traditional wave of quantum mechanics which are time-asymmetric, such as the *radiation* of the wave representing a photon emission, are not real, and are not represented in the wave form described by the new theory.

12.09 Before the advent of quantum theory, the only real, substantial constituents of the universe were supposed to be the particles from which matter is built, but the Nodal theory and the Copenhagen theory both question the reality of these particles. However, whereas Nodal theory claims that such particles exist only at *nodes*, and cease to exist between one node and the next in a particle's life history, several other interpretations question the reality of particles between one *observation* and the next. The consequences of this latter view take their most extreme form when observations or measurements of a system are separated by considerable intervals of time, and during these intervals all the possible ways the system could develop are said to exist simultaneously as a vast *superposition* of waveforms. Only when the system is next observed do the waves collapse, and one out of the many possible outcomes crystalises out.

Until this occurs, according to those who adopt this viewpoint, the particles are said not to have unique positions and momenta, but to be in a superposition of incompatible states, and can not be said to exist in the usual sense of the term.

12.10 The best known illustration of this principle is provided by the famous Schrodinger Cat paradox. This unfortunate creature is incarcerated in an opaque box, which also contains a flask of poisonous gas, and a trigger mechanism capable of breaking the flask to release the gas, thereby killing the poor animal. The trigger is actuated by a small radioactive source which has a 50% probability of releasing an alpha-particle within the next hour, and if this does happen the cat will die. The box is not opened until the end of the hour, and so the cat has a probability of one-half of being found dead when the lid is lifted. Now the paradox concerns the state of the cat during that hour. Because no observation can be made of the animal until the box is opened, the waveform describing the box and all its contents must be in a superposition of two states, in one of which the cat is alive, and in the other it is not. The cat itself must be in a superposition of a live state and a dead state.

12.11 We can understand why conventional quantum theory forces us to this bizarre conclusion, for Schrodinger's Cat illustrates on a large scale a principle which we see in a less dramatic form in the familiar two-slit experiment of Thomas Young. The two possible states of the cat correspond with the two routes a particle can follow through Young's apparatus. We would like to say that the cat must be either alive or dead, and that the particle goes through one slit or through the other, but this leads to a contradiction, for the probability of the particle reaching any particular point on the screen would then be the sum of the probabilities that it got there *via* the one slit or *via* the other. This would not give the interference effect we observe; we know we must not add probabilities here, but rather the complex wave functions corresponding to the two routes, which differ not only in magnitude but also in phase angle. If the phase angles differ by about $180°$, the superposition can result in a zero magnitude at certain points on the screen, and only thus can we explain the dark bands in the interference pattern. In the same way, we are told, it is possible that, when we open the Schrodinger box, interference effects could be observed between the waveforms resulting from the alpha-particle's creation and its non-creation. So we must not try to imagine the cat existing, in either a live or a dead state, until the box is opened, just as we must not visualise the particle passing through either the one slit or the other.

12.12 Now the Nodal theory does admit that interference effects can result from the coming together of two parts of a wave function which, since they separated, have passed through intermediate nodes on their journey. We discussed two examples in Chapters 6 and 7, one relating to Young's experiment when the particles collided with low-energy photons, and the other when photons are reflected from a pair of mirrors. From the Nodal point of view, the only link between one node and the next is *information*, and in cases where all the necessary information can be passed from one segment of a particle's world-line to the next via a node, then interference effects can still follow. This is what happens in the case of the particle collisions and the mirror reflections, and in principle could arise when the lid of the Schrodinger box is opened. In practice, though, this would never occur, for any coherent information must certainly be lost when the flask is shattered; this is Bohr's "irreversible amplification", the moment at which he believed the waveform to collapse. Only later writers have tried to delay this moment of collapse until an observation is made, but in the Nodal interpretation neither of these moments has any special significance. Schrodinger's cat seems irrelevant to believers in the Nodal doctrine.

12.13 The Heisenberg uncertainty principle is easily explained by the new theory, but its implications are rather different from those arising in the traditional interpretation. The usual description of the position-momentum uncertainty of a particle tells us that the product of the uncertainties of these two quantities is of order h, Planck's constant, and assumes that both quantities have their own degree of uncertainty. But the fact that either can be measured as accurately as we wish by designing a suitable experiment, provided the other is not measured at the same time, led Bohr to suppose that it was meaningless to speak of such exact values except in relation to the apparatus used to measure them, and he sidestepped the issue of whether these values both had an objective existence if they were not measured. From our viewpoint a particle's position has no meaning between one node and the next, and so the question of accuracy of measurement does not arise. On the other hand we can say that the momentum has a precise value between one node and the next, for the wave corresponding to the particle, the NWF, does have an exact frequency. In principle we can measure the positions of nodes as accurately as our methods allow, but there is *no* direct way of measuring momentum. Any attempt to measure simultaneously the position of a node and the momentum with which the (imagined) particle approaches that node must involve apparatus to *change* the position of the node, for example by passing the particle through an aperture. It is not surprising that in these circumstances we must trade knowledge of the one against the other, and in simple cases the Heisenberg relation is easily verified.

12.14 What can we say about the *indeterminacy* of quantum phenomena? We know the future behaviour of a system is not determined uniquely by its past history, and so in general we cannot accurately predict the future by calculation, but to Nature herself this is of no importance. Our capacity to store information about the *past*, as memories in the brain or as records on paper or in computers, is one of the miracles of life on earth which may have no parallel anywhere else in the universe, but because our brains are thermodynamic mechanisms, driven by the inexorable entropy gradient without which we could not live, we will never be able to gain direct knowledge of the *future* in the same way. In the absence of life, Nature keeps no detailed records of either the past or the future, and so it is rather strange that we are puzzled more by our inability to know the future than by our remarkable capacity of knowing the past.

12.15 It follows that Einstein's belief in quantum systems' possessing some system of *hidden variables* of which we are ignorant, but which nevertheless determine their future, this belief also was unnecessary. There may indeed be processes at work in the micro-world of which we still know nothing, but if such new secrets are revealed in the future it seems unlikely they will change the way we view the unpredictability of quantum phenomena.

12.16 The Nodal Theory throws no new light upon the EPR experiments and the phenomenon of *entanglement*. Modern interpretations of these experiments all seem to offer no escape from the belief that the probabilities associated with entangled particles can be transmitted *non-locally*. By this we mean that the occurrence of one event can affect the probability of another even when the two are not within sub-luminal range. To observe these phenomena we must set up the same experiment many times, compare observations of entangled pairs of events, and assess the frequencies of related results. Then we find *correlations* which appear unlikely unless they are influenced by these non-local effects. It is only in this rather abstract way that the EPR results appear strange. But the concept of non-local influences is an essential part of the Nodal Theory in a much wider sense, for the Nodal Wave Form is "aware" of the layout of the surroundings of a system without restriction of space or time, and so perhaps the non-local conclusions of entanglement experiments are easier to accept, now that they are seen to be just one manifestation of a much wider principle. If we think such non-local influences are counter-intuitive, it is only because we never witness them in everyday life, and indeed can never be directly aware of them. The present theory reduces the behaviour of particles and radiation to relationships between nodes, and nothing else, but all the classical rules which we had come to believe such particles to obey are reflected in these nodal relationships. We still never observe particles, radiation or information travelling at speeds greater than that of light; it is just the transmission of *probabilities* which can occur at such speeds.

12.17 Much work remains to be done before the Nodal Theory is sufficiently robust to challenge all its alternatives. These chapters attempt only to sow the seeds, which now need cultivation. Perhaps there is here a fruitful field of research for young physicists seeking unexplored territory within which to work and make a worthwhile contribution to our understanding of the physical world. Were the author fifty years younger he would gladly lead the expedition, but he is now at the wrong end of his life to engage in such an adventure. What he can do is point out the directions in which it seems exploration should be profitable, to give Nodal Theory a wider base of knowledge and understanding from which to argue its case.

12.18 We took the first few tentative steps along this journey of exploration in Chapter 11. We tried there to separate those features of conventional quantum theory which are purely subjective from those which correspond to features of the real world, and which the

Nodal Theory attempts to represent. We decided that, among the objective characteristics of the Nodal Wave Function, its *frequency* must be real, for it bears a simple relationship to the rest mass or energy of the particle it represents. And the orientation in space-time of its wave-fronts determine the reference frame in which the particle is at rest and, together with the energy, decides the momentum which the particle will display in any other frame. But we decided that several aspects of the CWF must be purely subjective, either because they depend on our own velocity relative to the particle's world-line, and so upon the Einsteinian reference frame we happen to occupy, or because they reflect the imperfect knowledge we can have of a particle's present state or future history. In the first category we place the *wavelength* of the waveform we observe, for this depends wholly on the relative velocity of the particle's nodes and ourselves. We used the NWF to calculate the wavelength associated with a particle, as observed in the reference frame of any particular observer, and obtained the well-known de Broglie formula $\lambda = h/mv$. Into the second category we must place any feature of the wave description which is time-asymmetric, and also any which reflect our own ignorance, such as the spread in the values of λ and v when a particle is represented by a wave-packet, and various manifestations of the Heisenberg uncertainty relation. Again using only our definition of the NWF, we derived the Heisenberg relation between our knowledge of the position and momentum of a particle.

12.19 We discovered that a sort of probability significance can be attributed to the Nodal wave. Interference effects are certainly real and independent of any observers; when Young's experiment is set up there is no doubt parts of the screen lie in dark bands and parts in light ones, indicating clearly that the probability of any particular photon reaching certain points of the screen varies from place to place. When we make the hypothetical supposition that the time and place of the particle's starting node, (r_1,t_1), is fixed, while its ending node, (r_2,t_2), can take a range of values, we can consider the density distribution of points (r_1,t_1) if the experiment is repeated many times. With these conditions we find that the appropriate frequency distribution is given by $|\Psi|^2$ where Ψ describes here the NWF, just as it does for the CWF.

12.20 Finally in Chapter 11 we considered briefly the NWF for massless particles such as the photon, and we derived the correct relativistic formula for the Doppler shift in wavelength when a source of light and an observer are approaching or receding from each other.

12.21 We still do not know the nature of the nodes from which our world is built, nor of the NWF which connects them. We can picture the nodes as points where information is exchanged, like miniature telephone exchanges, and we can imagine the waves to resemble the electromagnetic variety with which we are more familiar, but it is unlikely that these images come close to the things themselves. Perhaps we shall never improve on such pictures, for we are dealing here with entities so unlike those of the familiar world, and our sensory perceptions of it, that our representations may always remain no more than mere analogies. But there are aspects of the present picture which are obviously incomplete, and we must hope someday to be able to add the missing features. We have, for example, some understanding of how the NWF transfers information about energy and momentum, but we cannot say the same about *electric charge* and *particle spin*, and those other characteristics of nuclear particles which are conserved at the interactions taking place at nodes. How does the wave represent these features, and transfer them from node to node?

12.22 The NWF, as we have described it, serves two functions. Firstly it carries information, such as the mass of the particle it represents and the orientation in space-time of its world-line between pairs of nodes. Secondly it helps determine the position of these nodes; for example points which are inaccessible because of intervening matter or interference effects are indicated by the low intensity of the NWF there. But in both respects our descriptions are incomplete, for in many situations the intrinsic *spin* of a particle should feature in both these functions. The spin of an electron or the state of polarisation of a photon is sometimes preserved through a series of nodes, showing that the relevant information is passed on from node to node, just as is information about energy and momentum. And in certain experiments, such as those involving polarisation, the spin of the particles can result in some destinations being less likely than others. The necessary information must be carried by the NWF, just as in interference experiments it is phase information carried by the wave which gives rise to light and dark bands on the screen or photographic plate. These considerations show that the NWF we have described in previous chapters needs to be supplemented in order to incorporate this spin information. This supplementary information must share two essential features with the NWF that we have already developed, namely its independence of any observer and its time-reversibility. Further work is required to derive the most plausible form for the new wave, and to show how it can explain spin and polarisation phenomena within the same limitations as our previous work has explained momentum and interference effects. This is perhaps the most pressing problem for NWF theory in the days ahead.

12.23 Indeed we do not understand properly what this property of spin is in itself. We may have in the back of the mind a picture of a spinning top, but we know the spin of an electron cannot really be like this. For a body to spin, different parts of it must be moving in different directions. So if an electron has no component parts, it cannot spin in this sense. Nodal theory provides some relief from this dilemma, for the electron does not really exist between its nodes, and the spin is seen to be no more than a part of the *message* transferred from one node to the next by the wave function. But there are experiments in which the combined spins of a large number of particles can produce a macroscopic change of angular momentum. How can a nonsubstantial wave impart angular momentum to a material body? Equally puzzling, why is the angular momentum always transferred in integral multiples of $h/2$?

12.24 May we end on a more abstract note? The purpose of the Nodal Theory, as of all the other theories of quantum mechanics, is to provide explanations of the phenomena we observe at the atomic and nuclear scale, to provide understanding. What do we mean by *explanation* or *understanding*? These concepts clearly differ from *knowledge* or *skill*, which give us the tools we use with notable success and without controversy in our work in sub-atomic science. Acquiring understanding seems to be more difficult than acquiring knowledge. One can have a detailed knowledge of the topography of a country without understanding the geological processes which have formed it. One can be a skilful car driver without being able to explain the chemical and mechanical operations which enable the car to move.

12.25 It seems that the explanation of a fact involves a higher degree of abstraction, or generality, than the fact itself. One must ask if this process of generalisation can go on for ever, whether we are pursuing an infinite regress, or whether we will someday reach the end of the quest, a single overriding principle which will explain everything, and give us an understanding of all the processes which drive the universe. Many a parent of a precocious child must have pondered on this when confronted by an endless sequence of "why?" questions. "Daddy, why is it getting dark?" "Because the sun is setting." "But Daddy, why does the sun set?" "Because it goes below the horizon." "Why?" "Because the earth is turning on its axis." "But what keeps it turning?" At this stage most parents will admit defeat. Sadly some will reply, "Oh, do stop asking silly questions!"; only those blessed with the same childlike curiosity as their youngsters, and faith in their growing intellectual grasp of facts and principles, may answer, "I don't know, but perhaps someday you will, and will be able to explain it to *me*."

12.26 Some philosophers have expressed the view that mere scientists will never reach the ultimate truth that answers every "Why" question, and that only philosophy will reach this final point.

> If one says to the physicist, "Now please tell me what exactly is energy? And what are the foundations of this mathematics you're using all the time?" it is no discredit to him that he cannot answer. These questions are not his province. At this point he hands over to the philosopher. Science makes an unsurpassed contribution to our understanding of what it is that we seek an ultimate explanation of, but it cannot itself be that ultimate explanation, because it explains phenomena in terms which it then leaves unexplained. (*Confessions of a Philosopher*, Brian Magee, Weidenfeld, 1997)

Of course it does, and so must philosophy! The only way to explain phenomena in terms which are *not* left unexplained at the end of the line is to explain them in terms which already *are* explained, resulting in a circular chain of reasoning which achieves nothing. Science's quest for the truth will never end, but this should not be a cause for regret. Indeed it would be a sad day if science ever did complete its task; the excitement of discovering, however slowly, the secrets of this amazing universe far outway the temporary satisfaction which would greet a final closing of the book of scientific progress. An old Taoist proverb says, "The journey is the reward". Perhaps our attempts to resolve the mysteries and contradictions of quantum mechanics will prove to be one of the most arduous stages of the journey, but any small step in this difficult terrain, if it proves to be a step in the right direction, should bring us not just a little more understanding, but also pleasure and satisfaction.

THE END

Appendix 1:
Nodal Wave Function of a Free Particle

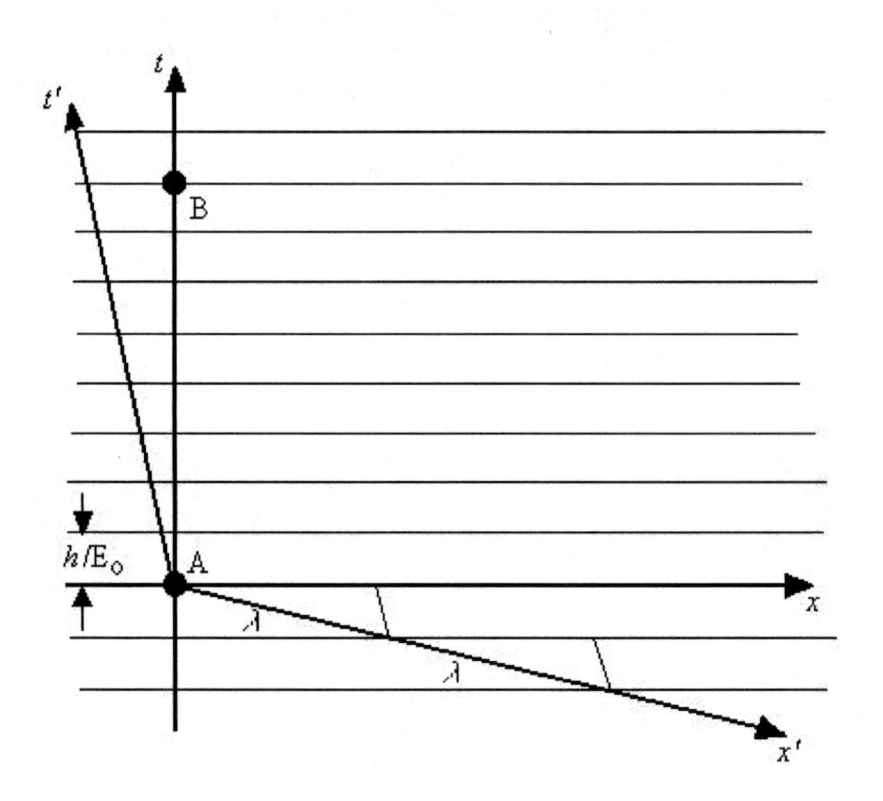

- 162 -

Assume the NWF of a particle with rest-energy E_o, in the reference frame in which it is at rest, to be:

$$\Psi(x, t) = a \exp(-2\pi i E_o t / h)$$

Let (x', t') define the reference frame of an observer moving with speed $-v$ along the x-axis.
The Lorentz transformation gives:

$$t = \frac{t' - vx'/c^2}{\sqrt{(1 - v^2/c^2)}}$$

So $\Psi(x, t) = a \exp\left(\frac{-2\pi i E_o (t' - vx'/c^2)}{h\sqrt{(1 - v^2/c^2)}}\right)$

Now if E is the total energy including kinetic energy in the (x', t') frame, and m is the particle's mass in the (x', t') frame, Einstein's relativity transformations give:

$$E = E_o / \sqrt{(1 - v^2/c^2)} \quad \text{and} \quad E = mc^2$$

So $\Psi(x, t) = a \exp(-2\pi i E(t' - vx'/c^2) / h)$
$= a \exp(2\pi i (mc^2 vx'/hc^2 - Et'/h))$
$= a \exp(2\pi i (px' - Et')/h)$ —————— A

where p and E are respectively the momentum and energy in the observer's frame.
This is the standard wave equation.

The wavelength $\lambda = h/p$.
Speed of wave along x' axis $= E/p = mc^2/mv = c^2/v$.

*****

Appendix 2:
The NWF of a Photon and the Doppler Wavelength Shift

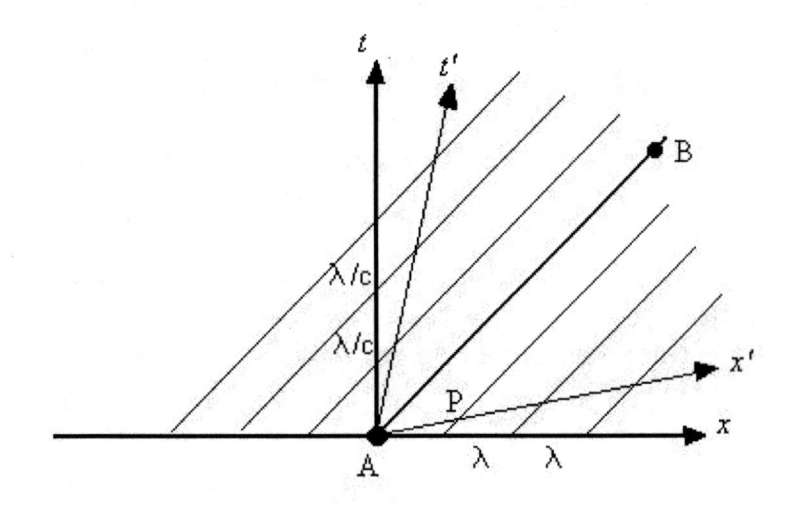

To find the NWF of a photon we *assume* Formula **A** from Appendix 1:

$$\Psi(r, t) = a \exp(2\pi i (px - Et)/h)$$

We know $E = hc/\lambda$ and $p = h/\lambda$ in this frame.

Thus $\Psi(r, t) = a \exp(2\pi i (x - ct)/\lambda)$ throughout all space-time. This is the NWF of the photon in this particular reference frame.

We now draw (x', t') axes for an observer moving along the positive x-axis with speed v.

To find the x and t co-ordinates of P:

Since $t' = 0$, $\quad t = vx/c^2 \quad$ (Lorentz transformation)

P lies on the first wave-front, so $\quad x - ct = \lambda$

Solving, $x = \lambda c/(c - v)$ and $t = \lambda v/c(c - v)$

Now $x' = \dfrac{x - vt}{\sqrt{1 - v^2/c^2}} = \dfrac{\lambda c/(c-v) - \lambda v^2/c(c-v)}{\sqrt{1 - v^2/c^2}}$

$\qquad\qquad\qquad = \lambda \dfrac{\sqrt{(c+v)}}{\sqrt{(c-v)}}$

This is the correct relativistic formula for Doppler Shift of Wavelength for recession velocity v.

*****

Appendix 3: Quantum Computing

App3.01 Much thought has been given in recent years to the design of the next generation of digital computers, and to the possibility of building machines which make direct use of the *quantum* properties of elementary particles. Some of the work currently being done in this field has direct relevence to the validity of the nodal hypothesis, and so we must enquire how it impinges on nodal theory, whether its discoveries might suggest modifications to the theory, or whether indeed the nodal hypothesis could become untenable if this work eventually succeeds.

App3.02 If such a quantum computer can ever be built, it will have some obvious advantages over today's machines. The last twenty years have seen a steady and remarkable increase in the power of digital computer systems; both the memory capacity and the processor speed of the small personal computers which can be bought for between £1000 and £2000 have practically *doubled* every two years, resulting in an increase of both these characteristics over the twenty year period by a factor of about one thousand. Alongside the increasing speed there has been a steady advance in the miniaturisation of components, so that today's computers are still about the same size as those of twenty years ago, despite their enormously increased power. These two attributes, size and speed, are of course related, and every increase in clock speed sees a decrease in the distance which signals can move, even at the speed of light, during one clock cycle. At the time of writing, clock speeds in the region of 1 GHz are common, so that in the duration of one cycle a signal can move no further than about 30 cm, and parts of a machine which are synchronised to the clock-pulses must lie within a very few centimetres of each other. We can expect this process of miniaturisation and increasing speed to continue for several more years, but we shall eventually reach a natural limit where the number of electrons comprising the signal currents will be too small to allow reliable operation. Clearly this barrier can be breached if individual molecules, atoms or sub-atomic particles themselves become the devices which store and operate on the bits of data which a computation requires to be handled, and this promise of smaller and faster devices provides one incentive for the development of components which function at the atomic level.

App3.03 An essential building block of the conventional computer memory is the *bi-stable*, a tiny electrical circuit that at any moment can display only one of two clearly defined states, and so can store a 0 or a 1, as required in handling binary arithmetic. In today's computers a *register* often consists of 32 bi-stables, and so the whole register can store in binary form any integer between 0 and $2^{32} - 1$, a total of about four billion possible integers. At least in theory, the world of atomic particles can supply us with unlimited "bi-stables", for many elementary particles possess an intrinsic "spin" and a magnetic moment, and whenever an attempt is made to determine whether the spin axis lies along a particular direction the answer is always just "yes" or "no". Under certain conditions a single particle can preserve its spin between one measurement and the next, and so can act as a ready-made bi-stable for storing a binary digit. The possibility of building a quantum computer using single elementary particles in place of electronic bi-stable circuits is currently being investigated at several research centres around the world.

App3.04 But what has caused the greatest excitement is the belief held by some physicists that a quantum computer, in addition to its increased speed and storage capacity, would permit the processing of vast amounts of data *in parallel* and *simultaneously*. In a traditional computer, when large numbers of data items are to undergo the same operation, as in many arithmetic procedures, this operation must be repeated successively as many times as necessary. If speed is vitally important, the only alternative is to use a large number of separate processors working simultaneously in parallel. Many physicists believe, however, that a quantum system can exist in a *superposition* of states, and does not collapse onto a recognisable state unless it is *observed* or *measured*. So whereas a traditional computer memory register can hold only one number at a time, a quantum register could hold simultaneously all the numbers of which it is capable. Furthermore, all these numbers can be processed together in one operation, provided we do not observe it until the task is completed. This parallelism is far more powerful than may be apparent at first sight. A conventional 32-bit register can hold at any moment of time just one of the four billion numbers of which it is capable, but if this could be implemented by 32 two-state quantum units, then between one measurement and the next this quantum register could be in a superposition of all its four billion states, and could store simultaneously four billion different numbers. If these all required the same operation performed on them this could be implemented instantaneously in parallel. It is easy to understand the enthusiasm of those currently working in the field of quantum computing.

App3.05 For many mundane purposes the speed of such a quantum computer would be of little advantage, for it would not provide a noticeable improvement over a conventional machine. But some apparently simple tasks seem incapable of solution on today's computers in a reasonable time because of the vast number of repetitive operations needed. If you are asked to add together two ten-digit numbers with paper and pencil, this will require just ten additions. If you need to *multiply* together two such numbers you will have to do one hundred multiplications. We say the *complexity* of adding together two *m*-digit numbers is of *order m*, and of multiplying together two *m*-digit numbers is of *order* m^2. In both cases the process can be performed by a computer in *polynomial time*, for m and m^2 are both polynomials. But some calculations cannot be performed in polynomial time; their complexity may be, for example, of order 10^m, which for sufficiently large m will require a much greater time than *any* polynomial function you can write down. Such processes are said to need *exponential time*, and it is for this class of calculation that the quantum computer, if it can ever be built, would prove invaluable, for it could reduce some exponential time problems to polynomial time.

App3.06 One notable example of such a problem is the *factorisation* of large numbers. Without using your calculator or computer, can you find the factors of 9,881,383 ? However good your paper-and-pencil arithmetic, this problem would probably occupy you for several days; factorisation requires exponential time. But interestingly the reverse problem, of forming a number when its factors are known, requires only polynomial time, and is far less complex. Try multiplying together 2657 and 3719. This should take you no more than about one minute, and the product is, in fact, 9,881,383. No doubt a conventional computer could factorise this number in a second or two, but one with a hundred digits rather than seven might take it hundreds of years. The virtual impossibility of factorising large numbers is the basis of some modern cryptography. The codes that are used to transmit sensitive financial data often rely upon this fact, and would be vulnerable if a new generation of computer were able to accomplish such factorisations in a reasonable time.

App3.07 Never has the gulf between theory and practice been so wide. At the time of writing (2002) very little practical work has been carried out that can reasonably be called "quantum computing". One approach is described in a paper by Matthias Steffen *et.al.* of Stanford University, (*Toward Quantum Computation*, IEEE, 2001). They have developed a substance whose molecules include five fluorine atoms, and each atomic nucleus displays one unit of spin. They dissolved this substance in a liquid and placed a sample in a Nuclear Magnetic Resonance machine. By subjecting these molecules to magnetic pulses of carefully chosen frequency and duration they were able to control and measure the spins of the five fluorine nuclei individually, and they claim to have created a computer with two registers, one of two bits and one of three, capable of holding any integer from 0 to 7. (They were, of course, manipulating a large number of molecules in parallel. But this is not the essential parallelism of quantum computing, which was represented by the superposition of states of the atomic nuclei in each individual molecule.)

App3.08 Experimental work is being performed elsewhere, but as yet it is all on this same minimal scale. This contrasts with the considerable amount of theoretical work being done at many different centres around the world, devising tasks suitable for quantum computer solution, designing on paper the necessary logic gates, devising possible computer architectures, and writing code for the actual programs to run on quantum computers which as yet exist only in the imagination. Some of these tasks will demand registers comprising hundreds or thousands of quantum two-state units, and no-one knows of what these will consist, or whether they will work.

It is acknowledged that a major hurdle to overcome will be "decoherence", the collapse of a particle's wave form whenever it reacts in any way with its surroundings, and the consequent loss of any information it may be carrying, whether or not this is part of a "superposition" of data. But there will undoubtedly be many other problems, some of which have not yet been conceived.

App3.09 A significant advance was made in 1994 when Peter W. Shor published details of an algorithm which he claims will enable quantum computers to factorise numbers in *polynomial* time, and so permit the factorisation of numbers far larger than any traditional computers can manage (*Polynomial-time Algorithms for Prime Factorisation*, IEEE Computer Society Press). The machine will require two quantum registers, which we shall call A and B, and the total number of bi-stables needed will be no more than about ten times the number of digits in the number we wish to factorise, which we call n. The number in register A at any moment we shall call a, and that in B is called b. The process can be divided into three parts, which we describe in some detail in an attempt to assess its feasibility.

App3.10 Firstly each element of register A is put into an equal superposition of its two possible states, so that the register as a whole is in an equal superposition of every possible number it can contain. As explained earlier, this could be a vast number; if n has about 80 digits, the number of possible values of a would exceed the number of particles in the universe, but these are all supposed to exist simultaneously in superposition. A small number x is then chosen almost at random, and Register A is subjected to a quantum process which puts into register B the following function of the number a:
$b = x^a \pmod{n}$

App3.11 For the non-mathematician a simple example will show what this means. Let us take $n = 91$ (obviously ridiculously small, but sufficient for illustration), and $x = 3$. Each value of b is the *remainder* when x^a is divided by n, and the following table shows this for the first few values of a.

a	x^a	b
0	1	1
1	3	3
2	9	9
3	27	27
4	81	81
5	243	61
6	729	1
7	2187	3
8	6561	9

It will be seen that the values of b are forming a sequence which repeats itself after every 6 values, and this continues for all values of *a*. We call 6 the *period* of the sequence, and denote it by *r*. In a real application of the method, *r* might be a very large number. Shor explains that register B then contains a superposition of all these values of *b*, just as A contains a superposition of all possible values of *a*. What we want to know is the value of *r*, the period of the *b* sequence, for it is easy to find at least one factor of *n* if the value of *r* is known (provided *r* is an *even* number; if not we can try again with a different value for *x*).

If you want the mathematics:
r is the first non-zero value of *a* for which $x^r = 1 \pmod{n}$
So $x^r = mn + 1$ where *m* is some whole number.
So $x^r - 1 = mn$
So $(x^{r/2} - 1)(x^{r/2} + 1) = mn$
It follows that at least one of $(x^{r/2} - 1)$ and $(x^{r/2} + 1)$ must have a factor in common with *n*.

App3.12 If we now *observe* register B, standard quantum theory decrees that its contents must *collapse* at random to one of its possible values (in our little example, perhaps to 9). But furthermore, because the registers are *entangled* by the quantum process which calculates the *b* values, register A must collapse also, but only to those values which make this *b* value possible, and so *a* now contains a superposition of this restricted set of *a* values (in our example, 2, 8, 14, 20, etc.).

App3.13 In the example it is clear on inspection that the value of *r* is 6, but if *r* lay in the millions or billions its value would not be so obvious. So Shor now applies another process to the superposition of *a* values, and finds their *discrete Fourier transform*, a process that can be done in a quantum computer using a method devised by D. Coppersmith (*An approximate Fourier transform*, IBM Research Report RC19642 (1994)). In effect this transformation replaces the (restricted) set of *a* values by their *spectrum* in the same way that a spectroscope analyses light of different wavelengths into a spectrum of bright coloured lines, or an acoustical spectrum analyser splits up a sound wave into its component harmonics. If we now *observe* the A register it will collapse to indicate a randomly selected one of the "harmonics", and if we repeat the whole process several times and record the values we obtain, it will be clear that all are multiples of some fundamental frequency, from which we can find the period of the original *b* values, or *r*. This gives us at least one of the factors of *n* as explained above, and we can then reduce the complexity of the problem by dividing *n* by these factors. It may now be possible to find the other factors using a conventional computer. If not, we can employ once again our quantum computer to reduce the problem a stage further, until we have found all the factors of *n*.

App3.14 Now what contribution can the Nodal Theory make to this new discipline? It will certainly ask for some of the speculative precedures being investigated by workers in this field to be *described* in somewhat different terms from those at present in use. But will it help to assess whether quantum computers can ever be built, and if so how useful they are likely to be?

App3.15 The essence of all these quantum methods lies in the manipulation of wave functions of superpositions, such as that described above. This manipulation must necessarily remain hidden from view, for conventional quantum theory tells us that the wave function of a system *collapses* to a single state whenever we observe that system. But Nodal theory adopts a different view. It maintains that these wave functions have *no real physical existence*, for they must belong to the class we have called *conventional wave functions* (CWF) and not to the class of *nodal wave functions* (NWF), as defined in Chapter 5. This classification is demanded by the strong time-asymmetry of the collapsing waves; any NWF must be totally time-symmetric. Thus the CWF represents only the constructions we ourselves *create* in our attempts to predict the future behaviour of a system which we cannot know for certain because our knowledge is always restricted to the past. When we describe a system as being in a superposition of different states, and assign quantum amplitudes to each state, nodal theory asserts that these amplitudes merely indicate the (complex) *probability* that such a state will be found to occur when we subsequently observe the outcome, a probability that may well be personal, in the sense that different people with different knowledge of the system's past history can quite properly assign different amplitudes to the same state. Only one outcome actually occurs, and to nature herself all the others mean nothing; they are significant to us just because in our ignorance we believe they might possibly come about. So when we describe a system as being in a superposition of states, each with its own probability of being realised, the system is really in just one of those states, moving towards the outcome which eventually is realised. All the other states represented in the superposition exist only in our minds.

App3.16 Let us examine the first stage of the Shor process from the two viewpoints, that of quantum theory as conceived by Shor and his colleagues, and that of nodal theory. The binary number a, contained in Register A, is acted upon by the function-generating hardware that calculates the value of x^a(mod n) , and this number is stored in register B. The conventional view is that A is in a superposition of states in which it contains every one of the numbers of which it is capable, these are simultaneously processed by the function generator, and then register B is in a superposition of states corresponding to all the numbers that can result from such a calculation. Then when we look at B its waveform collapses and we observe just one value of b, chosen at random from these possible values. What is the Nodal description of the process? In its original state, Register A contains just *one* value a, and this one number is processed by the function-generator and put into register B. Then when we look at register B we see just this one value of b, chosen at random from the possible values. We see that the conventional viewpoint and the nodal viewpoint describe the process very differently, but both appear to describe the observed outcome equally well.

App3.17 But in the next stage of Shor's process the nodal theory scores the advantage, for the conventional view has to explain how, when we observe B and collapse its wave-form, this collapse is somehow transmitted backwards into register A. A is supposed still to contain a superposition of a values, but only that reduced set which are consistent with the value of b which we observed in register B. On the other hand the nodal viewpoint has nothing to explain, for A only ever contained one value, and this value was responsible for the number in B which we observed.

App3.18 The third stage, the replacement of a in register A by the result of applying a Fourier transform to a, shows up the real difference between the two theories. According to Shor's theory A contains a superposed set of values for a, and the Fourier transformation should reveal a sort of "line spectrum" for this repetitive series, from which the period r of the series can be found. But the nodal theory insists that A can contain only *one* value; applying Fourier to this does not achieve very much, but if the process worked at all it would generate a *continuous* spectrum. When we come to observe the final value in the A register we are as likely to obtain any one value as any other; it will tell us nothing. So Nodal theory predicts that Shor's method, despite its ingenuity, will not work.

App3.19 What exactly does this mean? At first sight it may seem that, when we build our first machine to apply the Shor method, we shall be able to distinguish between the two theories. If the machine works then conventional quantum theory is vindicated. If it does not then it is condemned, and the nodal theory, while not proven, is still in the running. But the situation is not as simple as this. We have maintained throughout this treatise that the Nodal theory does not tell us anything about the results of experiments or observations; it is concerned only with reasons and interpretations. So the success or failure of a quantum computer can neither support nor refute nodal principles; it can not decide between nodal theory and mainstream quantum theory.

App3.20 The writer believes, however, that Shor's method, along with the work of several others doing research in the field of quantum computing, do not represent mainstream quantum thinking. The essential feature of all proposed schemes is the possibility of putting a system into a superposition of states, and of preserving this superposition through a number of processes. The avowed belief seems to be that unless someone *observes* the system then the waveform does not collapse. Perhaps the most familiar example of such a system is provided in the well-known legend of Schrodinger's cat, which we mentioned in Chapter 12. When the moment comes for the fateful nucleus to divide, or not to divide, as the case may be, it is put into a superposition of two states, with very different consequences. Then these two chains of consequences are supposed to co-exist until the moment the box is opened. If no-one looks inside for one hour, then the dead cat and the live cat must have co-existed in superposition for this length of time. Now the writer does not think a majority of physicists really believe this, and they are in good company, for it was Niels Bohr himself who declared that the collapse of the waveform occurs not just when someone "looks", but as soon as any "irreversible amplification" takes place. The shattering of a flask of poison is certainly irreversible, and is also an amplification of the splitting of a single atom, so the superposition must have collapsed long before the cat dies (or not, as the case may be). The nodal theory is more specific, and declares that in general the waveform of a particle collapses as soon as it reaches its next node. (The only exceptions are in situations like reflection from a rigid mirror, when the wave gives up only a small part of the information it carries, and is able to go on to the following node sufficiently intact to cause interference effects there, as described in Chapter 7.)

App3.21 In the final stage of any successful implementation of Shor's algorithm the set of numbers superposed in the A register is subject to a Fourier transformation. Shor summarises this process by quoting the mathematical operator which gives the final state of the A register in terms of its previous state, as follows (where q is the total number of possible a values):

$$|a\rangle = \frac{1}{q}\sum_{a=0}^{q-1}\sum_{c=0}^{q-1}\exp(2\pi iac/q)|c\rangle$$

App3.22 It will be seen that each of the superposed numbers must be multiplied by every other number in the superposition. This requires each bit of each number being accessed a large number of times, sometimes being found to contain a "1" and sometimes a "0", and this information is used to influence that part of the system which is summing the results of these multiplications. It seems certain that both mainstream quantum theory and the nodal theory deny the possibility of this happening. Standard quantum theory acknowledges that a particle in a superposition of states can give up its information once only, after which it will give the same value each time it is measured. For example, a photon which has been transmitted by a polaroid filter will be transmitted by any other such filter whose axis is parallel to that of the first. Both mainstream quantum theory and the nodal theory thus declare that such a quantum computer, relying on the simultaneous processing of a superposition of states, will not perform as required. Only the more extreme interpretations of Shor and his colleagues provide any basis for a belief that it will.

App3.23 So what are our conclusions? If such a parallel quantum computer is ever built, and if it ever functions as intended, then these extreme beliefs will be supported, and the future for computing will be exciting indeed. The traditional quantum theory of Bohr, Feynman and their conservative colleagues will have to be discounted, and the Nodal theory also must be consigned to the flames. The most secure coding systems used by the world's banks for transmitting sensitive information will become assailable, and a large part of the literature on quantum theory will have to be rewritten. But the writer is confident that this will not come about. Those presently engaged in the quest for a new species of quantum super-computer, for which they must believe that quantum superpositions can survive despite their internal interactions, will be shown to have backed the wrong horse. Mainstream quantum theory and the Nodal theory will remain as viable alternatives, and will be free to compete on the only playing field that is open to them, the field of plausibility.

*****

Bibliography

Bohr, *Discussions with Einstein*, Living Philosophers (1949)
Brody, T., *The Philosophy behind Physics*, Springer-Verlag (1993)
Chapman, M.S.(et.al.), *Photon Scattering from Atoms*, Physical Review Letters, Vol.75, No.21 (1995)
Coveney, P. & Highfield, R., *The Arrow of Time*, Allen (1990)
Cramer, J.G., *The Transactional Interpretation of Quantum Mechanics*, Reviews of Modern Physics, 58, 647-688 (1986)
Davies, P.C.W., *About Time*, Viking (1995)
Davies, P.C.W., *The Cosmic Blueprint*, Unwin (1987)
Davies, P.C.W. & Brown, J.R. (ed.), *The Ghost in the Atom*, CUP (1986)
Davies, P.C.W., *Other Worlds*, Penguin (1980)
Davies, P.C.W., *Space and Time in the Modern Universe*, CUP (1977)
Davies, P.C.W. & Betts, D.S., *Quantum Mechanics*, Chapman & Hall (1994)
Dirac, P.A.M., *The Principles of Quantum Mechanics*, Clarendon Press (1930)
Einstein, A., *Relativity*, Methuen (1954)
Einstein, A., Podolsky, B. & Rosen, N., *The EPR Experiment*, Physical Review 47, 777-780 (1935)
Feynman, R., *The Character of Physical Law*, MIT (1967)
Feynman, R., *QED*, Penguin (1985)
Feynman, R., *Lectures on Physics, Book III*, Addison-Wesley (1965)
French, A.P. & Taylor, E.F., *An Introduction to Quantum Physics*, Chapman & Hall (1979)
Gautreau, R. & Savin, W., *Modern Physics*, McGraw-Hill (1978)
Gribbin, J., *In Search of Schrodinger's Cat*, Corgi (1984)
Heisenberg, *The Copenhagen Interpretation of Quantum Theory*, Physics and Philosophy (1958)
Magee, B., *Confessions of a Philosopher*, Weidenfeld & Nicolson
Parr, H.C., *Time, Science and Philosophy*, Lutterworth (1997)
Penrose, R., *The Emperor's New Mind*, OUP (1989)

Penrose, R., *The Large, the Small and the Human Mind*, CUP (1997)
Polkinghorne, J.C., *The Quantum World*, Penguin (1990)
Price, H., *Time's Arrow and Archimedes' Point*, OUP (1996)
Savitt, S.F. (ed.), *Time's Arrows Today*, CUP (1995)
Schulman, L.S., *Time's Arrows and Quantum Measurement*, CUP (1997)
Sudbery, T., *Illuminating Entanglement*, Nature, Vol.379, P.403 (1996)

Index

Advanced action, 4.21
Advanced waves, Ch.8., 4.19, 4.20, 5.12
Angular momentum, 9.03
Annihilation, 5.02
Bell's Inequality, 9.06, 9.07
Bertrand,J., 3.12
Big bang, 2.10, 8.01
Bohm,D., 9.03
Bohr,N., 1.05, 1.07, 1.08, 3.03, 4.08, 4.10, 4.32, 7.03, 7.05, 7.09, 9.02
Boltzmann,L., 2.09, 2.20, 2.27
Boundary conditions, 2.11, 2.12, 8.01
Brody,T., 9.13
CWF, 5.15, 5.24, 8.14, 11.03
Carnap,R., 3.06
Change, 2.02, 2.05
Coherence, 6.11
Collapse of wave function, 1.04, 1.06, 1.09, 1.10. 1.16, App3.12, 4.07, 4.09, 4.10, 4.12, 4.17, 4.22, 5.12, 5.26
Collision, 1.11, 1.12, 1.18, 6.08
Complex number, 3.30, 4.05, 6.04
Compton effect, 4.03, 7.03
Consciousness, 6.06
Conservation of energy, 5.05, 5.09, 5.21, 7.03
Conservation of momentum, 5.05, 5.06, 5.09, 5.19, 5.21, 7.03
Conventional wave function, 5.15, 11.01
Copenhagen, 1.05
Copenhagen Interpretation, 1.05, 1.06, 1.07, 4.10, 4.13, 4.19
Counterfactual, 9.12
Cramer,J., 4.19, 4.20
Davies,P., 1.07, 1.09, 3.21, 6.06, 8.06, 9.04, 9.05, 12.06
de Broglie,L., 4.04, 4.06, 4.23, 6.03, 11.05
Decoherence, App3.08
de Finetti,B., 3.08
Determinism, 3.03, 3.29, 4.07, 4.10, 4.13, 4.14, 4.15, 5.19, 5.20, 12.14
Deutch,D., 4.12
Diffraction, 1.11, 4.04, 5.08, 5.23, 5.24, 6.03, 7.09,

>   7.10
Dirac,P.A.M., 1.14, 1.15, 4.23
Dispersion, 2.12
Dissipation, 2.12, 2.27, 8.04, 8.11, 8.12
Doppler effect, 11.15, App.2
EPR experiment, 3.28, 5.08, Ch.9, 12.16
Einstein,A., 1.03, 1.05, 2.04, 3.03, 3.28, 4.03, 4.08,
       4.13, 4.15, 4.15, 4.21, 4.23, 7.03, 7.05, 7.09,
       9.01, 9.02, 10.13
Electric charge, 12.21
Electromagnetic wave, 1.04
Electron, 6.04
Energy, 1.11, 4.02, 10.14
Entanglement, 12.16
Entropy, 2.07, 2.09, 2.10, 2.11, 2.27, 8.01, 8.11
Equation, 8.05
Ether, 10.06
Everett,H., 1.09, 1.16
Expansion of Universe, 2.10
Exponential time, App3.05
Factorisation, App3.06, App3.09
Feynman,R., 1.03, 1.11, 1.14, 5.10, 5.10, 5.11, 6.04,
       6.11, 8.06, 8.09
Flow of time, 2.05
Four dimensional picture, 2.02, 2.03, 2.21, 5.16, 7.06
Fourier transform, 11.08, App3.13, App3.18, App3.21
Free will, 2.19
Future, 1.19, 2.06, 2.15, 2.19, 2.24, 4.15
Galileo,G., 12.02
Gravity, 2.07, 2.10
Gribbin,J., 1.07, 6.06
H-Theorem, 2.09
Heisenberg,W., 1.03, 1.06, 4.23, 5.13
Heisenberg Uncertainty Principle, 1.11, 4.06, 4.07,
       5.04,   5.06, 5.22, 5.23, 6.09, 7.07, 12.13
Hidden variable, 3.01, 3.03, 3.11, 4.07, 4.10, 4.13,
       4.15,   4.21, 12.15
Imaginary number, 3.30
Imagination, 4.23, 5.16
Information, 1.13, 1.18, 4.20, 5.04, 5.05, 5.09, 5.21,
       6.15, 6.13, 6.16, 7.12, 12.12
Interference, 1.14, 3.02, 4.04, 4.08, 5.25, Ch.6, 6.04,
       6.11, 11.14
Irreversible amplification, 1.06, 4.10, 12.12, App3.20
Joint measurability, 9.13
Keynes,M., 3.06
Knowledge, 1.06, 1.09
Laser, 1.01
Lewis,D., 9.12

Light, 1.04, 8.11
Locality, 9.10, 9.15
Lorentz contraction, 10.07
Mass, 10.14, 11.21
Maxwell,J.C., 2.09, 8.15
Measurement, 1.04, 1.10, 1.17, 4.09, 4.21
Memory, 2.14, 2.18, 2.19, 4.15
Michelson,A., 10.06
Mirror, 7.12
Momentum, 1.11, 5.06, 5.18, 6.08, 6.12, 6.13, Ch.7, 10.14
Morley,E., 10.06
Multiplication of probabilities, 3.32
NWF, 5.15, 5.21, 5.24, 6.13, 6.14, 8.14, 8.15, 11.07, 11.10
NWF of atom, 11.16
NWF of free particle, 11.04, App.1
NWF of photon, 11.12
Newton,I., 6.01, 12.02
Nodal Wave Function, 5.15, Ch.11
Node, 1.13, 1.18, Ch.5, Ch.6, Ch.7, Ch.11, Ch.12
Non-locality, 3.03, 4.15, 4.19, 5.08, 5.27, 9.04, 9.15, 9.16, 12.16
Now, 2.03, 2.06, 2.06, 2.15, 2.17, 2.18, 2.20, 2.21, 5.16, 12.05, 12.06
Observation, 1.04, 12.09
Observer, 1.06, 1.07
Occultation, 10.06
Pair production, 5.02
Parallel computation, App3.04
Parallel universes, 4.12
Particle, 1.04, 1.05, 1.14, 6.14, 12.08
Partridge,B., 8.07
Past, 1.19, 2.06, 2.15, 2.19, 2.24, 4.15
Penrose,R., 4.14, 4.16
Personalism, 3.08
Phase, 3.30, 6.04, 6.10
Photoelectric effect, 4.03
Photon, 1.04, 1.17, 4.03, 4.08, 5.25, 6.11, 6.14
Planck,M., 1.11, 4.02, 8.15
Planck's Constant, 1.11, 4.02
Podolsky,B., 9.01
Polarisation, 1.17, 3.02, 9.03, 9.08
Polaroid, 1.17
Polkinghorne,J.C., 5.10
Polynomial time, App3.05
Price,H., 4.21, 8.02, 8.05, 8.08, 8.09, 8.10, 12.06
Principle of Indifference, 3.07
Principle of Relativity, 10.13

Probability, 1.14, 1.17, 2.12, Ch.3, 5.03, 6.04, 6.10,
    9.15, 11.11
Probability amplitude, 3.30, 4.16, 6.10, 8.13, 11.02
Probability density, 4.05
Probability function, 1.06, 1.07, 5.13
Psi, 5.03, 5.12
Quantum amplitude, Ch.3, 3.30
Quantum computing, App.3
Radiation, 5.12, 7.05, 7.08, 8.02, 8.13, 12.08
Range theory, 3.12
Reality, 1.05, 1.07, 4.11, 4.14, 6.16, 9.02, 9.10, 9.15
Records, 2.13, 2.18
Reflection, 7.12
Relativity, 2.02, 2.04, 2.24, 3.03, 5.09, 9.15, Ch.10,
    11.05
Retarded wave, 4.19, 4.20, 5.12, Ch.8
Reversibility, 2.07, 2.08, 8.01
Rosen,N., 9.01
Sample space, 3.11, 9.11
Schrodinger,E., 4.05, 4.23
Schrodinger's cat, 12.10, 12.11, App3.20
Schrodinger's equation, 1.07, 4.05, 4.09, 4.14, 5.10,
    6.04, 8.15
Schulman,L.S., 4.22
Second Law of Thermodynamics, 1.16, 2.07, 2.08, 2.10,
    2.11, 2.12, 2.15, 2.20, 8.04, 8.12
Shor,P.W., App3.09
Simultaneity, 5.08, 10.11
Space-time, 2.02, 2.11, 2.23, 2.28, 4.18, 5.04, 5.18,
    12.05
Spin, 9.03, 12.21, 12.22, 12.23
Sum over histories, 5.10, 5.11
Superposition, 1.09, 1.14, 12.09, App3.04
Thermodynamics, 2.07
Time, Ch.2, 4.15
Time dilation, 10.07
Time reversal, 2.26, 2.28
Time symmetry, 1.16, 2.08, 2.12, 4.17, 4.19, 4.21,
    5.02, 5.26, 7.08, 8.02, 8.09
Time travel, 2.23
Transactional interpretation, 4.19
Transistor, 1.01
Two-slits experiment, 1.14, 1.15, 5.25, 6.01, 6.03,
    6.04
Uncertainty, see Heisenberg
Velocity, 7.02
Velocity of light, 10.03
Venn,J., 3.14
Virtual photons, 11.17

Wave, 1.04, 1.05, 1.14, 12.08

Wave function, 1.07, 1.09, 1.10, 1.13, 1.18, 4.16,
    4.17, 5.03, 5.12, 5.14, 6.10, 7.02
Wave packet, 4.06, 11.01, 11.02, 11.08
Wavelength, 4.04, 6.09, 11.09
Wheeler,J., 1.07, 4.11, 6.06, 8.06, 8.09
Wigner,E., 1.07, 4.11
World-line, 2.03, 2.11
Young,T., 6.01

*****